Adam's Task

ALSO BY VICKI HEARNE

ADAM'S TASK

Calling Animals by Name

Vicki Hearne

Introduction by Donald McCaig

Skyhorse Publishing

Copyright © 2007 by Skyhorse Publishing, Inc.
Introduction Copyright © 2007 Donald McCaig

All Rights Reserved. No part of this book may be reproduced in any manner without the express written consent of the publisher, except in the case of brief excerpts
in critical reviews or articles. All inquiries should be addressed to: Skyhorse Publishing, 555 Eighth Avenue, Suite 903, New York, NY 10018.

www.skyhorsepublishing.com

10 9 8 7 6 5 4 3 2 1

Library of Congress Cataloging-in-Publication Data

Hearne, Vicki, 1946-2001
  Adam's task : calling animals by name / Vicki Hearne.
     p. cm.
  Originally published: New York : Knopf, 1986.
  Includes index.
  ISBN-13: 978-1-60239-002-7 (pbk. : alk. paper)
  ISBN-10: 1-60239-002-9 (pbk. : alk. paper)
  1. Human-animal communication. 2. Animal training. 3. Domestic animals--
Behavior. I. Title.

QL776.H43 2007
636.08'35--dc22

2006038125

Printed in Canada

FOR DICK KOEHLER

who taught me how to say "Fetch!"

and in memory of Bill Koehler

Contents

And out of the ground the Lord God formed every beast of the field and every fowl of the air and brought them to Adam to see what he would call them; and whatsoever Adam called every living creature, that was the name thereof.

Genesis 2:19

What will complete the human work is, however, not one other but only all others.

STANLEY CAVELL, The Claim of Reason

INTRODUCTION

There are many good books; thrilling books are rare. In 1986, when I happened across Vicki Hearne's essay "Crazy Horses" in *The New Yorker*, I felt like some homesick exile startled by a voice singing brilliantly in my native tongue. I read Vicki's essay phrase by phrase, let her phrases flow into her sentences and then, long before Vicki concluded, I returned to the beginning to start reading afresh. This is not because Vicki Hearne is difficult—though she insists that you pay close attention—but because I didn't want her story to ever end. "Crazy Horses" is one chapter in *Adam's Task*, a book which is certainly the finest philosophical animal study of our generation, and I am beginning to think the best of the twentieth century.

Let me backtrack to 1986. I was and am a sheepdog trainer. I believe that training any dog to anything like his full capacity is an intricate, heartfelt, deeply intellectual undertaking which deepens the trainer's soul as surely as it satisfies the dog's. The conversation between trainer and dog is so subtle, dense, and satisfying that I have known great trainers whose ordinary human speech has atrophied. These brilliant linguists cannot explain what they do, and often cannot answer novice's questions because asking that particular question means the questioner can't understand a true answer.

When *Adam's Task* was published, our national dog discourse—apart from the exemplars and anecdotes working trainers tell each other—was dominated by behaviorists whose claims to understand mammalian learning were couched in language so ugly it makes my eyes water; ethnologists asserting that since dogs are descended from wolves, one can best study dogs by studying wolves, although they do wonder why—since the wolf is altogether a better character—anyone would want to study dogs in the first place; Cartesian zoologists with their radical disdain for objects of their study and animal rights pioneers, like Dr. Peter Singer, who, having confessed that he didn't know much about particular animals and wasn't especially fond of them, proceeded to develop complex theories of how we should interact with them.

Adam's Task came into this linguistic briar patch with the aplomb of a D-8 Cat. What Vicki did—and this is her great achievement—was translate the conversation great trainers have with great dogs and horses into language all of us can understand. She brought three extreme vocations to *Adam's Task*: philosopher, poet and animal trainer. Elsewhere Vicki has written that she is proudest of the last. That surprising revelation tells us much, I think, about her seriousness.

The first readers to respond strongly to *Adam's Task* were eminent academic philosophers who loved her reasoning and were fascinated by the unusual subject that summoned it and workaday trainers who read the book for the stories of great dogs and great horses. *Adam's Task* made others mad as hell. Animal rights aficionados couldn't decide whether it would be better to simply denounce Vicki or co-opt her as "an animal rightist herself —if she'd only admit it!" Dog fanciers (dog show people) whose arcane lingo obscures and excludes these sparks found Vicki's pellucid prose "difficult" and her democratic spirit profoundly unsettling.

Years later, the debate roars on. Behaviorist training books still begin with impassioned defenses of the "misunderstood" B.F. Skinner, and dog fanciers natter on. But Vicki's thinking has profoundly and permanently altered the debate. *Adam's Task* is the intellectual foundation of how we are beginning to look at "domestic" animals—a looking which unconceals our mutual involvement, allows them and us our creaturely opacities, acuities and dignities.

I don't know how many times I've read *Adam's Task*. I do know that like all great books it speaks to me afresh and differently at each reading.

As I write this, I'm training two Border Collies for sheepdog trials and starting a three-year-old who has been a difficult family pet because her heart is too great for petdom. I tell her she can no longer employ the silly stratagems that have filled her empty hours, but to replace them I will show her a new coherence, the coherence for which she yearns. She hopes to do right, and her trust that I can help her find that coherence is her most poignant appeal. Without her hopes I could do nothing.

Failing that hope, or betraying it, is every trainer's greatest fear—to fail to bring this dog into coherence is to fail her soul and sully mine. Because dog training is such a peculiarly intellectual, spiritual endeavor, my preparation for a training session might include reading poetry or the psalms. Recently I've been rereading *Adam's Task*—it's a terrific mindset to bring to the training field where a young dog will shortly demand of you all you have. For the sheepdog trainer, at least, theoretical philosophy is a very practical discipline. Someone once asked the great sheepdog trainer, J.M. Wilson, if she should talk to her dogs. "Of course you should talk to your dogs, madam," Wilson replied, somewhat testily. "But you must talk sense."

In *Adam's Task*, Vicki Hearne teaches us how to talk sense.

—Donald McCaig
Yucatec Farm
Williamsville, Virginia

Preface to the 1994 Edition

In 1993 *Time* magazine announced that anthropomorphism is no longer a sin, that it's okay now to say that animals think, hope, are puzzled, have expectations, are disappointed, even, for some, make their own little plans in a time scheme of their own. That has happened since this book came out. Also, there have been a few wonderful books published—McCaig's work on Border Collies, Diana Cooper's *Night After Night* (about the Big Apple Circus), and something that marks a major moment, or discovery, of a possibly grown-up consciousness of animals, John Hollander's anthology, *The Naming and Blaming of Cats*. The idea of relationships between people and animals as a potential goldmine of speculation, indeed, of forms of life, is no longer so disreputable as when I was struggling for the understanding that became this book.

This cheers me. Even time cheers me. It is something, at nearing fifty, to find myself accompanied in what was, when I was groping toward it in the seventies and eighties, an eccentric, crank project—finding a language with which to reveal some of what seemed to me to be so crucial to the fact that good trainers, the ones whose animals are so confident and convincing at their work, are precisely the ones whose ways of talking violate the received precepts of religion and science. (They do this even when they also have the habit, when, as it were, wearing their Sunday best, of dutifully mouthing behaviorist, or, earlier, Catholic strictures.)

Yet there are a couple of things emerging from the eighties

that disquiet me. One is, in a way, trivial in this context, because it is merely a fact of history. That is the anti-dog movement, and the policing activities that go with it, which have become ferocious. The most visible aspect of this movement was expressed in the media as countless stories about how "vicious" pit bulls and other breeds are. Less visible is the fact that it is open season on dogs in general, and this phenomenon was sponsored and buttressed by anti-pit bull propaganda *coming from major humane organizations.* Hence, Britain has its Dangerous Dog Act, with the consequence that a lot of people don't celebrate Christmas anymore. As I write this, there is before the Connecticut legislature, and no doubt others, a Dangerous Animal Act that will make some of the mildest critters I know illegal in this state. Why this should be, why the rise of the animal rights movement and an increased interest in "humane" and "not for profit" activities should coincide with, and at times be indistinguishable from, relentless enforcement activities targeting dogs, is a topic for scholarship. All I want to note is that there is an enormous flow of mostly unexamined superstition about animals in this culture, that the twin images of the ferocious beast and the gentle, loving, free, or frolicsome creature are, if anything, more pervasive and influential than they were when I wrote this book.

Disturbing also is the divorce between training and the "new" behaviorists. It's disturbing in part because it means that there are dogs out there on drugs that needn't be, that could be dancing instead. It's mainly disturbing as evidence of the implacable distance that remains between various forms of knowledge. By the "new" behaviorism, I mean that board-certifiable specialty that has appeared in the veterinary profession. A lot of drugs are prescribed; this makes news because the drugs are L-Trytophan, Librium, Prozac—human drugs. This does not mean that animals are almost human, however, but rather that we are learning new dosages. When drugs are not prescribed, lower "octane" dog foods are, and spaying and neutering.

A friend's brother-in-law, a veterinarian, welcomes the new movement because he doesn't like putting healthy animals down for behavior problems (at the owner's request) and so is glad to be able to give them a pill instead. Since drugs of one sort or another are often a blessing, a momentary reversal of the Fall, this is not necessarily to be deplored, but it seems something of a shame anyhow, that the knowledge of dogs and of training doesn't—no, can't—make it over college walls. There is, despite the regular appearance in the *American Kennel Club Gazette* of a column by a "behaviorist," no genuine exchange between training and the academy. This is in part a function of the fact that trainers and veterinarians are now in competition for the same market, or in some cases think they are, so they fall to quarreling—at least the behaviorists fall to quarreling with some trainers. (Many trainers welcome the behaviorist move out of, I suspect, a willingness to deny the heart of what they are doing with their dogs.)

If you are willing to say that trainers have knowledge (tough for some scientists and philosophers), and that the scientists and even the new behaviorists have some sort of knowledge (that is tough for some trainers to acknowledge), then it looks as though they have knowledge of the "same thing"—that is, the behavior of animals, especially domestic animals. This is not so, no more so than it was when I was writing this book. The philosopher Stanley Cavell says that everyone turns from the world to a world; we are all, then, making reports from the field, and there are different fields. If a very serious dog and a very serious handler are lucky enough to walk into a serious world together, then there is, say, no biting problem. In a different world with a different handler, that's a different dog, and someone has to haul out either some doggie Prozac or the sodium phenobarbital.

A world. I can no more explain to most sheepdog handlers why I persist in obedience training Airedales, when I own a Border Collie, than I can answer the man at dinner who has never had an animal and wants to know why I train. This most obdurate of facts about human and animal existence, that we all occupy niches, say, is not to be altered by any wind of intellectual fashion: it will continue to take genius to acknowledge, well, not THE world, perhaps, but that there is THE world, which is to say, worlds beyond one's ken. Skepticism about animal minds is a kind of panic, whether the authority endorses or refutes anthropomorphism on the one hand, mechanomorphism on the other.

A major issue in this book is authority. Where does it come

from, besides our chimplike impulses? who has the right to command whom? and so on. Since I wrote, "We can command,
follow, only whom and what we can obey," meaning only whom
and what we can hear, respond to, I have been engaged in some
pretty active and sometimes hazardous battles, in court, in the
media, and elsewhere, in defense of dogs and people of one
description or another. If I were rewriting the chapter "How to
Say 'Fetch!'" that closes with the sentence I here quote about
commanding, following, I might add the word "coherently" after
"can." We can coherently command, follow, only whom and what
we can coherently obey. This is not to say that force and guile do
not produce many grotesque parodies of the relationship of
mutual respect and autonomy I am envisaging—only that to the
extent that we command what we cannot obey, we are engaged in
force and guile, not genuine authority. Such engagements are
inevitable, in the related but different ways death and taxes are; to
say this is not to gainsay the possibility of coherence, only to say
that it is temporal. This book is not about what a good thing
authority is, but about the taint in our authority.

On the whole, even though I have learned things in the last
ten years, I still believe this book.

Adam's Task

1

By Way of Explanation

The impulse behind this book is specifically philosophical, which is a way of saying that the circumstances of my life have been such that it mattered enormously to me to find an accurate way of talking about our relationships with domestic animals. It mattered to me as a dog and horse trainer for what I hope are obvious reasons. When you are incoherent in your notions about an animal you are working with, things do not go so well with the animal, and an animal trainer is a person who can't help but be uneasy about such a state of affairs, whether or not s/he has the linguistic wherewithal to articulate the problem and the solution properly.

If I had remained firmly within the worlds of discourse provided by the stable and the kennel, I might have been content, not because there is no philosophy in those worlds, but because there is such a rich and ever-changing web of philosophies when good trainers talk and write. These philosophies remember and speak to their sources in the thought of the past and are, unlike the general run of philosophies, continually tested and either reaffirmed or revised, since the world of the genuinely good dog or horse trainer is one in which reality is quite clearly, as Wallace Stevens had it, "an activity of the most august imagination."

However, my temperament regularly led me away from the kennel and tack room to university libraries and cafeterias, laboratories and classrooms. The result was that for some years I uneasily inhabited at least two completely different worlds of discourse, each using a group of languages that were intertranslatable—dog trainers can talk to horse trainers, and philosophers can talk to linguists and psychologists, but dog trainers and philosophers can't make much sense of each other. (Philosophers and linguists may have sometimes thought that they found each other incomprehensible, but their quarrels were usually about the interior decoration of the house of intellect and not about fundamental structural principles.) Because I had learned to talk, more or less, in both worlds, I was intensely alert to the implications of Wittgenstein's remark, "To imagine a language is to imagine a form of life."

Here is as good a place as any to speak of the example that most clearly indicates the problem I set out to deal with. In Germany there was once a cart horse named Hans, owned by one Herr von Osten. Hans had to back the cart he pulled in a circular drive, and his skill at doing this, the story goes, so impressed von Osten that he decided that horses in general and Hans in particular must be smarter than generally supposed. Von Osten began doing various things with Hans, teaching him to respond to questions either by tapping with a hoof a certain number of times or else by indicating one of a number of blocks on which the alphabet was written.

Hans was a good learner, and in time philosophers, linguists and psychologists from all over came to test his acumen. It turned out that Hans could not answer questions if he could not see the person asking him. It turned out further that if the questioner was in sight, Hans could always find out what the questioner thought was the correct answer, no matter how hard the questioner worked at remaining still and impassive. Hans apparently read minute changes in breathing, angles of the eyebrows, etc., with an accuracy we have trouble imagining.

This led to von Osten's being denounced as a fraud, and he seems to have died an unhappy man, not so much on his own account as on that of the horse in whom he so deeply believed. And there has now come to be a technical term in academic studies of animal psychology, the "Clever Hans fallacy." This is the fallacy of supposing that an animal "really" understands words or symbols when what the animal is doing is "merely" reading body language. In the literature, this notion is used to discredit virtually anyone who disagrees with the writer in question as either a fraud and a charlatan or else as just plain credulous and stupid. There is an unhealthy air of triumph in the rhythms of the prose of the people who do this discrediting, and I have found myself moved to wonder why, if the trainers and thinkers who believe that Hans illustrates something more important are so discountable, they must be so often attacked.

I told a friend of mine, the poet Josephine Miles, the story of Clever Hans. She said, in response to finding out that the humans couldn't conceal from Hans what counted as the correct answer, "But isn't that interesting!" One of the points of this book is to say, "Yes, Jo, that is interesting." She is now dead, so I can't say it to her, but I can say that she would probably want me to explain that, of course, when I here and throughout the book take swipes of one sort and another at academic thinking, they are lover's complaints—if I didn't love the worlds of discourse we call intellectual and academic, I wouldn't care if things went well there or not.

One of the worlds I lived in when I first set out to address this problem was the animal trainer's world—the trainer of domestic animals primarily, although that world is not to be located by the boundaries of kennels, racetracks, horse-show grounds or obedience trials. The other world was the world of the intellectual, especially the academic or full-time intellectual, though it is not strictly bounded by the walls of either university or editorial offices.

What happened was that in the mornings I would get up just before dawn and work my horses. Generally I had finished with most or all of them (it depended on how business was going or whether certain horses were giving me trouble) by noon, so I would shower and go over to the local university. There were a couple of people there I liked to meet and talk with over lunch, and I also liked to prowl around in the library and either take courses in or just hang around courses in phi-

losophy, psychology, zoology and linguistics. I had been bitten
in my childhood by a passion for books, especially books that
were, as a recent novel has it, "hard to read, books that could
devastate and transform your soul, and that had a kick like a
mule when you were finished with them." There were as many
glittering and lovely creatures in those books and in the conversations of people who cared about them as there were in the
kennel and the stable.

But despite their many beauties, most of the philosophers
and their associates in the libraries, and all but two or three of
the people at lunch, were profoundly disappointing, not in and
of themselves, but in terms of my passion for a language with
sufficient philosophical reach to tell me what I wanted to know
about the stable and the kennel. And there was a great deal that
tended to cause me to lose my temper, such as the enormous
amount of time that was spent in "curing" students and others
of saying precisely the sort of thing I wanted to say vigorously
and significantly about animals.

One thing that preoccupied me was the trainers' habit of
talking in highly anthropomorphic, morally loaded language.
That was the language I wanted to understand because it seemed
to me after a while that it was part of what enabled the good
trainers to do so much more than the academic psychologists
could in the way of eliciting interesting behavior from animals.
Trainers, for example, have no hesitation in talking about how
much a mare loves or worries about her foal, a cat her kittens
or a dog or a horse their work. But for philosophers and
psychologists to speak of love was to invoke abilities that are,
for reasons I am still not clear about, as rigidly restricted to
Homo sapiens as some religious doctrines have restricted the
possession of a soul to members of certain races, cultures and
sometimes genders.

In any event, the talk I heard was of no help in enabling
me even to figure out what my project was, though I knew a
lot after a while about what it wasn't. It wasn't behaviorism,
it wasn't ordinary-language philosophy and it wasn't classical
quantificational logic. Nonetheless, I saw many interesting
things along the way. A student giving a paper on post-par-

turition behavior in cats would inadvertently attribute to the mother cat a mental state, such as caring about her kittens. The student would be corrected and would learn in time to deliver solemnly quantified reports on the amount of licking behavior, suckling behavior and so on that was "exhibited" by the queens. I wondered about that word "exhibited." Exhibited to whom? The researchers? The kittens? I also wondered about the intellectual and spiritual futures of students so carefully instructed in the terrible grammar such ways of talking entailed.

Another habit that students had, curiously, to be cured of was the habit of supposing that one animal might hide from another animal. (I have never known a hunter to be successfully cured of this habit of mind.) I was deeply intrigued by this, for what in the world was the puppy doing under the bed when you returned home to find an unwelcome monument on the broadloom, if not hiding? But it was sternly pointed out to me what a great and anthropomorphic mistake it was to say or think this. In order to be hiding, whether from predators or from the vexed owner of the carpet, a creature would have to have certain logical concepts that animals simply couldn't have. I remember one careful exposition on the subject of octopuses, who will, in laboratory situations, hide behind glass in plain sight of predators. A number of things struck me about that seminar. One was the way the scientists cheerfully applied interpretations of the behavior of octopuses to the behavior of gazelles and St. Bernard puppies which seemed to me to demonstrate insufficient respect for the individuality of octopuses. Another was the indifference of the researchers to questions about the importance of vision for octopuses and their predators, and yet another set of considerations had to do with my reflecting that in the same position I would probably do the same thing, either out of mindless habit or because in the tanks in the laboratories there wasn't anything else but the glass to get behind.

But in order to hide, it was carefully explained, one had to have a concept of self. Not only that, one had to have the concept of self given by the ability to speak academic language, or at least a standard human language—a concept of self that depends on the ability to think. And, as one philosopher in-

formed me unequivocally, any sort of thinking requires "first order
logical quantification theory." Since I myself didn't seem, on investigation, to be using FOQ, I couldn't make much of this.

Since those days, certain conceptually laborious and interesting experiments involving gorillas and mirrors have weak-
ened the more rigid of the foundations of some of these
cognitive allegories, but there is still little help from science.
The work with gorillas seems to establish that gorillas share
with human beings a tendency, which Aristotle notes in the
opening pages of the *Metaphysics* and which Plato worked into
his parable of horse and rider, to rely on vision. Dogs and
wolves and other animals, by contrast, distinguish themselves
from other individuals, and friends from foes, by scent markers.
I don't know why one can't speak, at least tentatively and for
the sake of philosophical speculation, of a wolf's territorial
markings as being a series of scent mirrors, or, as fiction often
has it, signatures, and argue from that to a concept of self. But
I learned early on to be cautious about saying this sort of thing,
and I said less and less as time went on, except to the two or
three friends who were patient with my ramblings. My passion
to find a way to write about the language of people who actually
work interestingly with animals increased, however.

After trying to talk, I would leave the university in the
middle or late afternoons to work with a dog or so and any
horses that had been left out of the morning schedule. Here,
in the various training arenas, the discourse was radically different. It was, as I have said, anthropomorphic, "morally
loaded," as it has always been in the great training manuals.
By this I mean that implicit as well as explicit in the trainers'
language is the notion that animals are capable not only of
activities requiring "IQ"—a rather arid conception—but also of
a complex and delicate (though not infallible) moral understanding that is so inextricably a function of their relationships with
human beings that it may well be said to constitute those
relationships.*

---

*By "moral understanding" I mean that as far as a trainer is concerned, a
dog is perfectly capable of understanding that he ought not to pee on the bedpost

Xenophon speaks of horses "greatly appreciating" certain "courtesies," and, to the irritation of a more or less scientifically minded translator, of the "cunning" of certain hunting dogs in leading other dogs off the trail of a rabbit by barking or baying falsely. The editor and translator in question appends a footnote in which he indulgently explains and apologizes for Xenophon's naive little slip here in attributing such a degree of intellectual capacity for misdirection to a mere (helplessly sincere) animal. When I showed that passage to a friend of mine who is fond of fox hunting, he remarked rather gloomily, "I believe I know that darned hound!"

Xenophon wrote quite some time ago, but his notions and something like his language continue to echo in modern training, albeit revised, here and there expanded, here and there muted, as well as from time to time severely reduced. Trainers still speak of whether or not a horse is "mean," "sneaky," "kind" or "honest" and vary their approaches to situations accordingly, sometimes saying, "Hey! You've got to come down on that dog hard and fast and right now—that's a real hood." Or "Relax, there isn't a tricky bone in that horse's body; he'll take care of you." Or "Don't worry, he'll come around okay, he's no real criminal, just a juvenile delinquent." Or, in appreciative awe, "Look at that dog work. She knows her job, doesn't she?" Or, as a general principle of training, "But first and above all, the horse's *understanding* must be developed." Or "If you want to know where the track is, *ask your dog!*"

There seems to me to be something terribly important about this language and what it implies, partly of course because it is a language I myself speak, but also, as I began in time to notice in more and more detail, because one can *do* so much more with the trainers' language, despite the fact that in the mid to late twentieth century it sounds as it has for some time—at best naive and at worst offensive, somewhat in the way that *Huckleberry Finn* has sounded offensive to some. In the past, attempts

---

even though he might want to. Characterizing the dog's own formulations of this understanding is a separate matter. To say what I've just said is, of course, to make a claim about the nature of moral understanding.

to speak in the way I have in mind have been regarded as heretical as well as intellectually unsound. And the agitation expressed by some writers and thinkers in the face of the trainers' persistence in talking the way they do, as well as the uneasiness some trainers express in response to their awareness of the possibility of that agitation, and the attempts in the introductory portions of some training manuals to placate that agitation, suggest that modern injunctions against anthropomorphism have as much of a heretic-hounding impulse behind them as any of the older ones. When, for example, I gave a portion of the chapter "Tracking Dogs, Sensitive Horses and the Traces of Speech" as a talk at the New York Institute for the Humanities, one person in the audience said that what I was saying sounded a little, well, religious. I patiently worked at finding out what she meant by religious, and it turned out that she meant "anthropomorphic." I said, "Oh, yes indeed, that's the whole point of this project!" She wondered aloud if I should be allowed to teach in a university, and at a later talk, when I found myself seated next to her by accident, she asked me to leave the room. The morally loaded language of William Koehler's stunningly fine training books have led to any number of court cases and to one case of the books having been banned, for a while at least, in Arizona.

In academic opinion, the trainers are, not to put too fine a point on it, intellectually disloyal. This would not in and of itself be worth more than a few paragraphs of social history if having something to say about what animals are like—about the problem of animal consciousness—were not so ubiquitous a way of providing a rhetorical and conceptual frame for investigations of human consciousness in all sorts of areas. Whatever the author in question thinks women are like, or blacks, or philosophers, or Jews, or Republicans, or Americans, or whatever category defines the "we" of a given discussion, it must first be made clear that the "we" is to be distinguished from the animals. It generally takes no more than a paragraph or so to characterize all of animate creation that is not the "we" at hand, or it did until lately. Now there are respectable tomes that attempt to prove that animals feel pain and that this has consequences for human morality. And in response to this literature, usually called animal-rights literature, there are renewed versions of the claim that animals are absolutely different from the "we" at hand. When this enterprise began, I felt an upsurge of hope; surely a title such as *The Moral Status of Animals* would help me to expand my own project. It didn't, though, but there began to be, refreshingly, the occasional bit of common sense, as when Tom Regan points out that "if Professor Frey's dog is a normal dog, he will eat his lunch, and not his master."* This sort of thing was cheering, especially as philosophers like Frey are capable of quite extraordinary performances, such as the following:

> Now in the case of my dog, can anything like a ranking of rational desires be achieved?... When I put food before him, my dog eats it; when I throw the stick, he fetches it. Both he does unfailingly, unless he is distracted by some stronger impulse, such as, on occasion, sex; and in response to the question whether my dog desires or prefers eating to chasing sticks, I can only say he does both when the situations are to hand and no other impulse interferes. Several times, I have tried putting food before him and throwing a stick at the same time; each time he has sought neither the food nor the stick but stood looking at me.†

At first I thought this was some sort of irony, but it wasn't, it was just plain old lunacy and ignorance. Several of my friends, some of them philosophers and some of them dog trainers, refused to believe me when I told them about this and other passages in that book, and I had to show them the pages to maintain my credibility.

None of this had anything to do with the knowledge of training that I wanted to bring to bear on various questions, and the "new" philosophers of animal consciousness were no more interested in what the trainers had to say than more

---

*Tom Regan, *The Case for Animal Rights* (Berkeley: University of California Press, 1983).

†R. G. Frey, *Interests and Rights: The Case Against Animals* (New York: Oxford University Press, 1980), p. 137.

"traditional" writers had been. They found them just as vulgar
and heretical as the logicians and the church fathers had, and
they seemed even more aggressively unwilling to distinguish
between boar hunting for sport, the greed that builds appalling
feedlots for pigs and calves and high school dressage for horses.

The more of this sort of thing I became aware of, the more
ill-tempered I got. In my ill temper I began to notice a lot of
things that didn't quite amount to a philosophical ground for
honoring the trainers' anthropomorphic language but that I took
as license. I noticed that in obedience and riding classes, people
with training in the behavioral sciences hadn't much chance of
succeeding with their animals, and that the higher the degree
held by the person, the worse the job of training was likely to
be. And one of the reasons I was the audience for so many
lectures on the wrongness of the trainers' way of thinking and
talking was that the psychologists and philosophers had to bring
their animals to me because they couldn't housebreak them,
induce them to leave off chewing up the children or, in the
case of horses, get them to cross the shadow of a pole laid on
the ground. The trainers' dogs and horses, by contrast, would
move with courage and determination over difficult tracks and
obstacle courses.

The consequence of all of this was my being led to cast my
intellectual, literary and moral lot with the trainers, even the
sleaziest of them, despite my fondness for the wonderful creatures of philosophy and related disciplines. This didn't mean
that as a thinker I was free from the intellectual tradition I
inherited; like any other trainer of my time, I have been enriched
and bruised by what I might call "scientomorphism," by which
I mean Western faith in the beauties of doubt and refutation
that is one of our central intellectual virtues. And it is, in its
place, a virtue, but like any popular notion, it is rarely in its
place and tends to run amok and lead to the curiously superstitious notion that to have no reason to believe a proposition
is the same as having a reason to assert that the proposition is
false.*

*I discuss this more fully on pages 98–100.

I should confess that doubt ran amok in my own case even after I had worn out a number of bridles and leashes, and that for a long time, even as I became cranky with the philosophers, I tended to think of the trainers as skillful perhaps but philosophically naive. I hadn't noticed that genuine mastery of anything entails sound philosophical thought of one sort or another. When, for example, I read in William Koehler's book on guard-dog training about the importance of being sure that your prospective protection dog has a well-developed sense of "responsibility," I tended privately and only semi-consciously to think it was a pity that he didn't know better than to use such a vocabulary in relationship to animals. I managed to think that even though I already knew him to be one of the greatest animal teachers the world has ever known.*

It was not, finally, the trainers who showed me the necessity of believing them, but a dog and, later, a horse. In this my story is a common one. Alois Podhajsky, the famous trainer of the Lipizzaner Stallions at the Spanish Riding School in Vienna, calls his autobiographical book *My Horses, My Teachers*, the true title of the autobiography of virtually every horse trainer who ever lived.

The dog who forced me to notice what was going on was an Airedale Terrier named Gunner. I was working him on a scent problem, having him follow a track laid by my seven-year-old daughter Colleen. As I work on tracking, the dog is taught not only to follow a scent but to retrieve objects dropped by the track layer. The track was plainly marked for me, since there was still dew on the ground and Colleen's footsteps showed clearly. Furthermore, I knew where the track "had" to end, since Colleen had been picked up in a car and driven away after dropping the last glove. I knew that she hadn't been in the area the track was laid in for a week, so there was no

---

*He was, for example, for years head animal trainer at Walt Disney Studios, a genius at training scout dogs, war dogs and police dogs, and the author of the best-selling dog-training book of all time, *The Koehler Method of Dog Training* (New York: Howell Books, 1964). His son, Dick Koehler, is at least as fine a trainer as his father and the finest teacher of anything I have ever known.

problem about a confusion of trails. Suddenly Gunner abandoned the trail and began bounding to the left, toward some
bushes about eighty-five feet away. I decided, as humans tend
to, that I knew more than the dog about what was going on.
I shouted angrily and tried to halt him with pressure on the
harness, but he kept on merrily (he always looked merry), to
my intense aggravation, and emerged from the bushes with a
stuffed toy Colleen had been for some days mourning the loss
of. It took me a decade to figure out how to talk about training
in general and tracking in particular in a way that would make
it clear why at such a moment my intellectual loyalties shifted,
and how to tell other stories, especially a horse story, that
would indicate *what the trainers have in mind* when they talk the
way they do.* But the experience was an epiphany rather than
a demonstration for me, the moment when, taking the stuffed
toy from the joyous young Airedale, it dawned on me that
people like Koehler use terms like "responsible" in relationship
to animals because those are the terms that *make sense* of the
situation.

I began realizing other things as well, such as that in the
trainers' world different kinds of animals exist than the ones
that I heard and read about in the university. For the trainer
there are hot working Airedales, dutiful and reliable German
Shepherds, horses with intense, fiery and competitive temperaments, other horses who are irredeemably dishonest. In the
universities, there were more or less Cartesian creatures of
uncertain pedigree, revised by uncertain interpreters of Freud
and Jung, which may be why in the world of letters in general
animals are invoked to mark "primitive" and usually unsavory
impulses, while in the trainers' world they are more like characters in James Thurber, who insisted that dogs represent "intelligence and repose" in his work. The trainers' language was, if
I could only unfold its story with the full acceptance of what
Stanley Cavell has called "the daily burden of discourse," the
right language, the philosophically responsible language.

*A version of the horse story I am thinking of is given on pages 117–21.

Knowing this was important to me. It enriched my work and conversation as a trainer, but it didn't enable me to tell anyone else much about what I was at last beginning to have a grasp of. There was no philosophical prose rhythm available for me to ring the right sort of changes on. I was able to sketch some of what I had in mind relatively soon in verse, but that was thanks to the virtues of poetry itself, which has wings and is good at dodging, able, to use Stephen Dedalus' phrase, to "fly by the nets" cast by the shadows of dark philosophies, ideologies and bad poems. I was also able to get some expression of the matter into fiction, at least to my own satisfaction, largely, perhaps, because philosophy tends to ignore fiction. But I wanted it to be *philosophy*, or something very like philosophy.

This was a terrific problem for me because argument was at the center of most of the philosophy I knew, and I didn't want to argue with anyone. Then two events occurred quite close together. One was the publication of Stanley Cavell's *The Claim of Reason: Wittgenstein, Skepticism, Morality and Tragedy*, which not only gave into my keeping certain philosophers and problems more securely than I had ever had them before, but also opened the possibility of a prose that was sufficiently subtle, muscular and accurate for me to ride in quest of the meanings I still needed to catch the meanings that eluded me. (It takes a meaning to catch a meaning, as Robert Tragesser, a splendid philosopher, once remarked.)

The other event was the arrival in Riverside, California, a few months later of Washoe, Moja and Loulis, three signing chimpanzees, and my coming to observe them and to talk with two of the people who worked with them. This happened at a moment when the debates about whether or not what the chimpanzees were doing could be accepted as language were particularly hot, and I suddenly had an occasion to begin writing, after hundreds of false starts.

That is one reason why this book, which is primarily concerned with domestic animals, especially working animals, begins with speculations about a wild animal. Another reason may be that, like the thinkers I have complained about, I needed something against which to define my subject. For me, I suppose, human beings and working animals are the "we," so it was natural enough for me to define my territory against the background of wild animals, whose worlds are far more various than my gestures in their direction indicate. In any event, the book begins at the point where my small knowledge and my vast ignorance met. Most of the questions I encountered quickly became questions about language, questions that located the boundaries of language in regions often understood to be remote from language.

There is one more piece of autobiography I would like to insert. A year or so after Gunner had handed me the central revelation, but some years before the simultaneous arrivals of *The Claim of Reason* and Washoe, I met the poet John Hollander, and we talked about animals.

I told him that I thought that the training relationship was a moral one, and he asked me, "Why do you say that?"

I replied quite crossly, "Because I think it's *true!*"

Fortunately for me, he very gently responded, "That's a good answer, but what I meant was, 'What do you *have in mind* when you say that?' "

No one else had ever wanted to know, so I began trying to explain, and in one way this book is simply an extended attempt to answer his question. And it was his poem "Adam's Task," and the generosity of the poetic thought in that poem about what naming is, that gave me both my title and a portion of the intellectual energy I needed to work out how to write *Adam's Task*. So the book is for him, and for Dick Koehler, who taught me how to say "Fetch!" But it is in memory of Gunner, who was the one who so generously and vigorously brought matters before my conscious mind.

And it is for Donald Davie, who taught nobly at Stanford. Were it not for him and his poetry and what I learned from them about narrative, fighting fair and something that could be called discovering the textures of actuality, I might still be sitting in the back of the lecture hall, in mute frustration.

And for Eleanor Carey, Marsh Van Deusen and Robert Tragesser, quick and subtle listeners who were able to hear and respond to the bits of genuine thought that emerged from my ramblings as acutely as even Gunner ever did.

2

A Walk with Washoe: How Far Can We Go?

Washoe, the first of the chimpanzees to be taught Ameslan, the American Sign Language of the deaf, is not a domestic animal, not one of those animals whose nature or temperament is not only the result of working with human beings but also what makes working with domestic animals possible. She does not have what animal trainers call a working temperament. This book is primarily about animals that do have working temperaments, so Washoe is outside of its subject matter, and one of the reasons I begin outside of my subject is to make a survey of its boundaries, in order to get a better broad view. That is also why I am not so much interested in Washoe in particular as I am in Washoe-for-example. What she is an example of is a relationship between human beings and animals.

A great many people who have been involved even tangentially with the signing chimps have been troubled (as I have been). The chimps are compelling—the rush to the typewriters to report on them (even in cases like that of Chomsky, who never left his study in order to find out what he was reporting on) has been extraordinary. But so has the failure of true and unquarrelsome meditations on the phenomenon they represent.

The rush in the writings of some thinkers suggests that when Washoe signs, "Give Washoe drink," we face an intellectual emergency. It may be that any challenges to our tacit assumptions about where language is to be located are, in this century, emergencies in the way challenges to explicit or tacit assumptions about the nature and location of the soul used to be; or it may be just that any genuine philosophical problem is an emergency. But the kind of case that Washoe is an example of has sponsored more turmoil than is easily explicable. What in our assumptions about language, or animals, or exchanges, or relationships is threatened? I have come to believe that the question "Why is it upsetting?" is prior to and perhaps contains in its answer some of the answers to the question "Is it language?"

So I am rushing to my own typewriter to report on why a dog trainer is troubled. And I understand myself to be writing about Washoe's *training*. Some readers may find this offensive because the word "training" invokes for them what behaviorists do with levers and electrical shock, or what sadists do with their victims, and they may wish that I would say *teaching*. Others may find the word inappropriate if their experience with dogs and horses tells them that Washoe has not achieved the noble condition that the expression "well trained" implies for an animal trainer. Both objections are worthy. I want to say *training* because I feel that I can push through to a more satisfactory view of what is going on if I read the problem of Washoe, not as a puzzle, but as a training problem.

In the course of working out training problems and understanding them (this may come long after their resolution, if it comes), one sometimes has to solve puzzles. But the problem—the difficulty, that is—comes before the puzzle, even though it is sometimes a puzzle that signals the presence of a hitherto latent training problem.

Yvor Winters, in the introduction to *Forms of Discovery*, said that the most important difference between a chimpanzee and a professor of English is that the professor has a greater command of language. He says that the professor may fancy himself a handsome fellow, but the chimp thinks otherwise and is unarguably the better athlete. He adds that the chimpanzee has no way of understanding the nature of this difference between them. He goes on to remark that the most important difference between a professor of English and a great poet is that the poet has a greater command of language. He pictures a hierarchy of command, not unlike the spiritual hierarchy in some medieval and Renaissance world views, in which differences in degree of command become at some point differences in kind, differences in kind of command and kind of understanding. Command of language is, in the case of Washoe, what most of the discussion is about.

A dog trainer has different views and different interests; s/he is interested primarily in respect for language. Command of language is something that we understand imperfectly, largely because we understand command and commanding imperfectly. Our imperfect understanding is revealed for some of us by the fact of the signing chimpanzees, and by certain tangles in the discussions of them. That is the emergency. We suddenly feel that we don't know what we're talking about.

The ability to recognize command of language is deeply important in our ordinary lives. If I meet you, a stranger, on a deserted street and discover that you are competent in the forms of exchange familiar to me—the rituals of "Hot enough for you?" and so forth—I am less likely to worry about whether or not you are going to kill me than if you, say, fail to respond to my "Good afternoon," or respond in a way I don't recognize. If, on the other hand, I should get to know you and discover that you can speak very well indeed—are able to discuss the writings of my favorite moral philosophers with intelligence and wit—I may quite confidently invite you into my home. It is possible to make mistakes about people in this way, but in general speaking well elicits trust. We want our leaders to be able to give good speeches. This is so deep in us that we are bewildered when we discover that the professor may be a murderer, or that the Nazi can discourse beautifully on the music of Mozart. And we have still failed to come to terms with Ezra Pound's fascism.

Command of language is a clue we use with one another, but command of language turns out to be useless without respect for language. If I respect your words that means that I give myself to responding meaningfully to what you say—that I won't suddenly decide in the middle of a lunchtime conversation to withdraw or to scream you into a terrified silence so that I can grab your wallet. If we converse, it also possibly means that when you discover your wallet is empty, I will be happy to pay for your lunch. Talking entails care and care-taking. That is part of what respecting one another means. Other sorts of linguistic confrontation, such as marital battles and various forms of preaching and opining, are not *talking*. The syntax of them is not the syntax of what we have in mind when we say, "At last, someone to talk to!" If the syntax of our lunchtime conversation changes from talking to arguing or preaching, our relationship has altered, and we have changed position with respect to each other.

With dogs, the situation is similar. The better trained a dog is—which is to say, the greater his "vocabulary"—the more mutual trust there is, the more dog and human can rely on each other to behave responsibly. "Responsible" may seem an odd expression to use in reference to an animal, but it is the only term that makes sense of certain training situations. Lassie and Rin Tin Tin, with all of their unlikely heroics, are successful characters because they provide meaningful emblems of our relationships with dogs. There is a connection, too, between Lassie's cleverness—her ability to fetch slippers and carry messages—and her reliability when the going gets rough. The circus dog who spells things with alphabet blocks is the dog who is able and willing to advance on the villain in the face of gunfire at the climax. The same is true of horse stories—"intelligent" and "bold" are synonyms in discussions of Trigger or Tony the Wonder Horse or Two-Bits, the New York City police horse in Irving Crump's tale, in which, for the human heroes as well, "educated," "smart" and "courageous" are virtues that seem to entail one another.

In real life, the case of competent police-dog trackers indicates what I have in mind. A good police dog has not only a large vocabulary but also extraordinary social skills. He understands many forms of human culture and has his being within them. He can be taken to the scene of a liquor-store robbery and asked to search, with the handler trusting that he won't molest the customers or other police officers or the clerk behind the counter. He knows what belongs and what doesn't, sharing our community and our xenophobia as well. He can take down a criminal who is attacking his handler on Monday and on Tuesday play with the patients at the children's hospital. These dogs, then, are glorious, but for anyone familiar with working dogs they are not *surprising*, any more than your pet dog is surprising in his or her ability to distinguish between your friends and strangers.

But someone might say that a dog's courtesy with guests is surprising, or that it ought at least to be remarked on that such profound connections between two species can happen at all. (It should be surprising, perhaps, that we can talk, and, of course, some philosophers have been surprised.)

Consider, for example, what happens when you train a wolf, or what happens at least when I train a wolf.* The wolf, or coyote, may sit, heel, stay, come when called and so forth. But a wolf doesn't respect our language, and his behavior can be accounted for pretty well with a stimulus–response model, from our point of view if not from the wolf's. The wolf may also become fond of me in some fashion or another, but I can't use him as a guard dog. Not only will he not distinguish particularly between family, criminals and guests, he will not have the courage of a good dog, the courage that springs from the dog's commitments to the forms and significance of our domestic virtues. The wolf's xenophobia remains his own. With other wolves he may, of course, be respectful, noble, courageous and courteous. The wolf has wolfish social skills, but he has no human social skills,† which is why we say that a wolf is a wild

---

*Here, as elsewhere, when I say "wolf" I mean *Canis lupus*—the wolves of the North Americas. I. Lehr Brisbin has alerted me to the temperaments of Asian wolves, which I don't know firsthand but which I am given to understand "fit" much better into human societies and hearthsides than the North American wolf does. My only hands-on experience is of the latter.

†There are, of course, stories, such as Kipling's *Jungle Books*, of humans and wolves learning to live together. There is also the story of Tarzan of the Apes. I suspect that these stories are accurate in their revelations of wolfish or chimpish civilization, and also accurate in that it is always, to my knowledge, the human

animal. And since human beings have for all practical purposes no wolfish social skills, the wolf regards the human being as a wild animal, and the wolf is correct. He doesn't trust us, with perfectly good reason.

The wolf is not alone in his regard for the commitments talking with humans implies. Even Lucy, the chimpanzee whose (true) story is told in *Growing Up Human*, turns out on examination not to have learned from her family, the Temerlins, who brought her up as their "child," as much about not biting and toilet training as the family dog. At the end of the book, the author has discovered that he and his wife want "a more normal life," and while they reject the possibility of zoos and chimp colonies for their "daughter," the book closes with the variously interpretable assurance that "all I can say definitely at the moment is that part of the earnings from this book will be used to establish a trust fund for Lucy, to provide for her care and comfort throughout her life." I do not doubt the love the Temerlins have for Lucy—but it is not generally necessary to pension off the family dog for the sake of the marriage! And no account I know of concerning work with wild animals gives useful advice for dealing with the fact that wolves, lions, tigers, orangutans and chimpanzees remain willing to commit mayhem no matter how large their vocabularies. In order to know more of what this is about, I'd like to take another look at dogs.

First, though, another small reminder about respect. If you and I are talking together at lunch, and you suddenly leap up and run out of the room shouting, "Watch out!" I will, unless I have the impulse to discount you, assume that something has happened—that you are, despite the oddness of your behavior, a reasonable person, and that I ought to find out what I should be watching out for. If I decide that you have gone mad, or are tricking me, then we won't be able to talk about it, though we will be able to argue. Similarly, if a detective suddenly changes his or her behavior in the course of an investigation, his or her partner will, if the working relationship is based on

---

who has to learn the foreign language and culture. We probably are the best users of language.

respect, assume that s/he has reason to do so. Otherwise the
working relationship breaks down, or even ceases to exist.

I mention these examples to make clear what I mean when
I say that it is by the same token that, when a good tracking
dog turns left at the corner of Ninth and E streets, the handler
will respect his judgment even though witnesses have said that
the dotty gentleman who has escaped from the old people's
home turned right at that point. The handler will usually continue, respecting the dog's superior knowledge, or had better
do so, without worrying a great deal about what "respect" and
"superior knowledge" are. It is enough to know when respect
breaks down, and to know this is to know a great deal. With
horses, respect usually means respecting their *nervousness*, as in
tales of retreating armies on horseback traversing minefields, in
which the only riders who survive are the ones who gave their
horses their heads, or tales of police horses who snort anxiously
when a car in a traffic jam turns out to be carrying the thieves
who escaped capture six months earlier.

I don't mean that handling dogs or horses seriously means
living in a world where respect never breaks down—in that
world, as in an exclusively human world, the possibility of
discounting is the context within which respect has meaning.
Xenophon and every other writer in my ken who places the
relationship with dogs in a moral or metaphysical context
reports on the irritating, irresponsible tricks dogs play, like the
one of deliberately barking on the wrong trail. People who
deliberately lead each other astray are considered culpable
because it is assumed that they are capable of behaving well.
(Chimps are not assumed to be capable of behaving well.) And
dogs and horses, like doctors, teachers and judges, don't necessarily get out of it when carelessness or some other lapse in
concern is to blame rather than mischievousness or malice.

A trained dog, a dog with a vocabulary, is sane and trustworthy. And training or retraining a crazy dog—one that has
had, say, schizophrenic experiences with phony and bizarre
distortions of attack training, or one who has taken to biting
in a desperate attempt to interfere with a childish or hysterical
handler who expects the dog to "want to please"—is a matter

of teaching him to respond coherently and meaningfully to *what is said*. A dog who will respond to talk will stop biting and will not turn on his master even if (especially if, actually) the dog is a German Shepherd or a Doberman Pinscher. Dogs that "turn on their masters" have had relationships with human beings that are in many ways like the relationships some of the mentally ill have with parents who are overtly appalled and secretly delighted with hostile behavior. Such a parent can't teach anyone to talk.*

The moral transformation of the dog comes about through stories, stories that provide a form of life within which responding to what is said is a significant possibility. Dog trainers like to tell stories about their dogs and other dogs, stories that have a number of functions. One function is to probe—to prove — the relationship between human and dog in a way that reaffirms the personhood of each. The stories, if they are elaborate enough, are frequently about people in confusion who, through the shock that comes from recognizing the reality of the relationship with the dog and then through the development or the restoration of that relationship, are enabled to put their own moral and social world in order. The dog may, through an act of devotion or heroism, compel acknowledgment. Sometimes, in the middle of such stories, the relationship breaks down, and the entire world is thrown into confusion through the handler's or someone else's failure to be true to the integrity of the dog. The structure of such stories, though it varies in completeness and sophistication, is remarkably like the structure of *Our Mutual Friend*, in which, at the very center of the novel, all of London,

---

*When this chapter appeared in *Harper's Magazine*, my remarks about human craziness inspired a great deal of angry mail. The most frequent complaint was that it was cruel to the parents of the mentally ill to suggest that they had anything to do with the sufferings of their children; they suffered enough as it was without people talking the way I do. My position in the world remains nonetheless the animal trainer's position. That is to say, while I certainly don't recommend guilt to anyone, as I have never seen guilt do any good, I do believe that it behooves us as humans to be alert to opportunities to change crazy-making grammars, regardless of whether they are something we are to blame for or not. Such opportunities are not always forthcoming, of course.

and thus all of the cosmos, is in doubt and bewilderment when the hero is no longer visible as a moral center. If the dog is not a hero, then he may sometimes be a Shakespearean fool, ignored in the middle of tragic storms.

I'd like to use an example from a "true life" dog story. It concerns Rinnie and his handler, John Judge, who were the pride of the Wichita, Kansas Police Department. Rinnie's nose was foolproof, his heart gallant, brave and dedicated, his mind alert and questioning.

One night Chuck Smith, who had the job of collecting supermarket receipts and placing them in the night deposit at the bank, called the police to report that he had been kidnapped in his own car and robbed. The police asked Smith to take his car back to the point where the kidnapper had gotten out of it, and Rinnie and John Judge were dispatched to track down the villain. Rinnie, on arrival, was asked to search the car. After taking a good sniff, the dog, calmly and without hesitation, walked around the car to where the victim was talking with police officers, and bit him in the seat.

The comedy was lost on John Judge, who was flabbergasted and chagrined. He took Rinnie severely to task, and the dog was disgraced. While Smith was taken to the hospital, John Judge and Rinnie went back to headquarters. The news about Rinnie's mistake spread like wildfire and was featured by all of the news media: "Rinnie's misdeed was a welcome event for the anti-dog faction. Letters were dispatched to the chief and the mayor. A thorough investigation of the incident was requested, and Rinnie was suspended from the force. The Canine Corps was in jeopardy."

It is important to notice that the mistake was conceived of as an extraordinarily clumsy one, unworthy of "the most inexperienced police dog." This matters because it indicates the depth of the loss of faith, the darkness of soul, of the moment when John Judge reprimanded Rinnie. When a police dog bites a victim, the perdition of the handler is absolute. The center does not hold, things fall apart. The dog's potential for virtue, and for lapse, is greater than the policeman's for lapsing from human law.

Fortunately the story goes on. A minor character (one of the detectives who is not named in the account I read) delved into Smith's background and had Smith submit to a lie-detection test. The machine, like the dog, said that he was lying, and further investigation revealed that Smith and an accomplice had planned the robbery together. (The significance of the parallel between the machine and the dog, and of the fact that the machine's authority was higher than either John Judge's or Rinnie's, belongs to another discussion.) At the end of the story, John Judge and Rinnie are restored to honor, the criminals are in prison and order is restored to Wichita. Order is restored, that is, by the reaffirmation and acknowledgment, on the part of humanity, of the moral meaningfulness of the dog's actions. To assert this is, of course, to proceed rather blithely past looming philosophical and psycholinguistic questions. This is what the stories do for trainers, enabling them to dissolve problems instead of solving them, so that they can get on with their work with dogs, ignoring for the nonce vast territories of philosophy that began when Aristotle, in the *Nichomachean Ethics*, denied casually and in asides that animals (and women) could participate in what he called the moral life.

There is another, related sort of tale, one as deeply informative, about retrieving. In such stories dogs perform spectacular, even impossible retrieves that amount to a transfiguration of their predatory "instinct" (an odd term for a large collection of abilities, including keen observation and analysis). In a comic version involving betting and brandy, a great retrieving dog, hunting in downtown New York, returns triumphantly to his master with an expertly stuffed pheasant. In more serious versions, retrieving transfigures the world through exaltation, just as in actual training situations formal retrieving transforms predation into an exultant submission to form that is the basis for both joy and commitment, a kind of marriage of the quest and the hearth. The intractable Pointer Hardhead, for example, becomes a field-trial champion, his stubborn and wild ways transformed into glee and impulsion that keep him going harder, faster and more alertly than the competition. Addie May gets her wheelchair, Little Valentine's heart need not be broken and the entire order of the world is affirmed as that of a world in which life is not only possible but glorious for all concerned.

Such stories are repeated over and over, not only in fiction, but in the lives of the people who tell them. The dog who is brought to Rudd Weatherwax because he's "wild and uncontrollable" becomes the film star Lassie. (My use of the masculine pronoun is not accidental—the "Lassie" in question was a male named Lad.) The dog a desperate owner offers to Bill and Dick Koehler because he "bites everybody" becomes Duke, the spectacularly cooperative star of such films as *The Swiss Family Robinson* and later has the courage and nobility required to take it on himself to run interference between homicidal Brahma bulls and their fallen riders.

The trainers say in one fashion and another, "You've got to talk to your dog." I'd like to go on a bit longer about how talking changes the dog's hunting impulse. A good dog begins life with the "instinct," if you will, to hunt: that is to say, to take possession of the thing he chases, to claim it as his own. This, whether or not the word "instinct" is appropriate, is as primal and visionary a part of the dog as the erotic is for us, or the impulse to ask unanswerable metaphysical questions. But dogs and people, unlike wolves and people, have the impulse to "play fetch" with each other, and the impulse to play fetch is the best predictor of good working dogs. It tells you which of a litter of eight-week-old puppies is most likely to develop the sense of responsibility required of a good Guide Dog. (Wolves may love you, but they won't Fetch, and they are poor guides.)

The impulse to play fetch is also a pretty good predictor of which of a group of eight-year-old human beings is likely to make a dog trainer. Dogs are domesticated to, and into, us, and we are domesticated to, and into, them. The potential dog trainer, obeying both instinct and myth, picks up a stick and throws it for his or her new puppy. The first time, Fido is fairly likely to bring it all the way back. The second time, however, Fido typically says, "Well, this is fun and all, but can I trust her with *my stick?*" So Fido compromises by bringing the stick to a point just out of reach and dropping it there so that the human, if she wants to play fetch, must accept this modified version and pick up the stick herself. Thus begins a game that can be played until the dog or the owner dies. It is fun, but it will seem to anyone familiar with it that no power on earth will induce the dog to bring the stick the extra three feet or ten feet forward, a move that would amount to a full acknowledgment of the human as an authority.

In formal training, the dog is forced to come those extra three feet, and to present the dumbbell or the bird to the owner. Some dogs take more kindly to this than others, but all of them have their doubts about it, and the most enthusiastic ball-playing dog on the block may put up a surprisingly vigorous fuss in formal retrieving situations. This fuss is, of course, very different from the wolf's response. The wolf simply never sees the point, even if, through stubborn and hard-nosed conditioning, he is brought to go through something resembling the formal actions of retrieving.

The dog is a domestic animal, and the postures appropriate to his life with human beings come to transform him and the action he performs, even if it is done mechanically and reluctantly at first. If training is completed properly, the dog makes an intuitive leap—joins the group, as it were—and may later display degrees of ingenuity and courage in finding lost objects and lost children that astonish the uninitiated. The handler, too, changes through his acceptance of posture and responsibility. He joins the group, too, enters the moral life as well, and learns to talk to Fido. (A failure on the handler's part to submit as fully as the dog is asked to results in a travesty of the training relationship that leads to mostly but not entirely misguided comparisons between obedience work and Nazi Germany. But more of this later. The complexities of that issue are out of place here.) The coherence created by training accounts for why it sometimes happens that the drunk or the juvenile delinquent or the supposedly "autistic" adolescent will "reform" as a consequence of training a good dog. They learn how to talk meaningfully with the dog, and then they learn to talk to the dog trainer ("He digs holes in Ma's flower bed. What should I do?"). Finally, talking may become possible with almost anybody who is willing.

The story may go like this: The borderline schizophrenic, through luck and because he has read *Lad of Sunnybrook Farm* or *Big Red*, ends up in the class of a competent dog trainer. He plows through, more or less blindly, with a faith born of dimly remembered tales. The going will be quite rough, sweaty and frustrating, and he is likely to give up without remembering dimly the right portions of the tales, the portions about patience and so on, and there will be moments when the trainer will have occasion to say to him or to someone else in the class something restraining such as "Excuse me, sir, but the command is 'Fetch!' 'F-e-t-c-h.' Not 'Son of a bitch!' but 'Fetch!' "

Then one day when he says, "Joe, Fetch!" Joe does a real retrieve, a retrieve that could go through fire. This may be the first time in the handler's life that language has proved—probed—the world and drawn a full, meaningful and serious response from another being. He steps, for the moment at least, out of schizophrenia and into position next to his dog, a whole human being in that moment, though not necessarily from that moment on. He also, incidentally, steps out of the schizophrenia of American myths of the splendors of isolation. Blocked, frustrating, enraging and covertly or openly murderous transactions simply lose interest at this point. And if he happens to be around people who don't have their own schizophrenic interest in blocking language, he will learn to talk to them. He will come to tell his own stories, and he may win trophies, which is fun and which is also a trope of acknowledgment.

Dog and handler, having learned to talk, are now in the presence of and are commanded by love. (This will happen even to people who don't start out as borderline schizophrenics.) The dog's apparent command of human language may be limited, but his respect for language commands him now, with his handler, as deeply as only a few poets are commanded. In this sense, command of and by language and respect for language are one.

But, as I have said, there are deep frustrations in the training process. These come about because the ability to utter, "Joe, Sit!" creates the illusion that Joe can know thereby exactly who we are, that we can penetrate his otherness, that he can through the phrase alone share our vision of the Sit exercise. It is rather like what we may feel when we ask someone to scratch our back, and it turns out that asking by itself doesn't make it possible for one's friend to scratch one's back in precisely the right manner. Anger results, anger that is the brother or perhaps the father of murder. In the dog story, and in real dog training, language both creates and absolves, placates, that anger.

The poet's condition and the dog's is that through obedience to whatever condition of language happens to lie at hand, they can move for a while through flame, even the frozen flame of despair at the condition of language. Our condition—those of us who have not submitted to despair—is that we have sufficient respect for language, some of the time, to talk and to refrain from murder. What, though, is Washoe's condition? And what are the stories about her and about chimpanzees in general?

In my life there aren't any very good stories about chimpanzees. I do have stories about my dog, an Airedale, who used to lie on the floor resignedly waiting for me to be done with my typing, a coherent waiting born of the logic of the inheritance passed to him by dogs whose masters read Dickens and by the great nineteenth-century breeders in Britain and Europe who had new conceptions of the dog as citizen.

I don't have any tales that would enable me to train a chimp, but there are, of course, tales about wild animals. There is *The Yearling*, at the end of which the deer's maturity and wildness force the humans to return it to "nature," with a shotgun. There is Daniel in the lions' den, a tale of a rendezvous and not of a marriage. There is *The Fox and the Hound* (the book by Dan Mannix, not the pseudo-Disney movie), whose story, despite the fact that the fox is hand-raised, is about enmity between the fox and the man-dog hunting team, an enmity as passionate as enmity between mutually domestic creatures ever gets, which makes it curiously parallel in some ways to tragedy. There is Farley Mowat's *Never Cry Wolf*, in which watching wolves and yearning in some ways after their life leads the narrator to begin sleeping wolf-fashion, which he says causes his lover to leave him when he brings the habit back to civilization.

There are some very bad movies about chimpanzees living with people and dressing in human clothing, and there are, lately, real-life stories about chimps such as Lucy and Nim Chimsky living with families. The movies use preadolescent chimps and don't confront the issue of what to do with sexually mature ones. The stories about Lucy and Nim Chimsky are stories about the ultimate unworkability of living with chimpanzees. Also, they tend to be clogged with more or less freudian (I don't mean that they sound as though Freud had written them—very little that is "freudian" does) analyses that, for me, make the most sentimental of stories in the tradition of *Lassie Come Home* seem like rooms full of intellectual freedom, light and air. They certainly have nothing of the serious trainer's philosophical toughness about them. There is also *The Talking Ape*, but while I find that Keith Lardlaw is much more grown-up in his descriptions and love of his orangutan than the authors of most of the other stories I've seen, *The Talking Ape* ends with the ape in a zoo, which is not a horrible fate as Lardlaw describes it, but which is still not my notion of a training story.

There are stories, but none is of much use to me, so I had no tales to take to Gentle Jungle, the wild-animal training facility where I found Washoe, together with her adopted son Loulis and another female chimp named Moja. Washoe has not always lived in a cage, but caged she was when I saw her. While I am not automatically moved to pity by the sight of a cage, this nonetheless affects profoundly the possibilities of description and narration available to me, since I have no story, no paradigm, and must resort to anecdotes and journal entries. I am virtually alone in front of Washoe's cage.

It is seven a.m. I am with a friend, on the grass, under a tree in the main compound of Gentle Jungle, an organization that rents trained wild animals to movies, television and so on. The main compound is an area about the size of a football field, ringed round with cages that contain Bengal tigers, Galápagos tortoises, pumas, baboons, a wolf, spider monkeys, various sorts of bear. These are wild animals. I don't know how to talk to them, and as an animal trainer I feel anxious about this.

My friend is a linguist and a philosopher by inclination. He is here to find out whether or not Washoe "has" language. I have discovered that that question causes a kind of hot fuzziness in my head and have left it aside for the moment. I am hoping to find out what Washoe's story is.

Roger Fouts, who has done much of the significant work with Washoe, arrives and starts signing with her and with Moja. My friend asks me, "Are they talking? Is Washoe talking?"

I reply, "I don't know, I haven't met her." As it turns out, I won't meet her, or at least I won't do what I have in mind when I report that I have "met" a dog or a horse or a human being.

It occurs to me that it is surprising that "I don't know, I haven't met her" is rarely the response given to "Can Washoe talk?" If I ask you whether or not Fred Smith can talk, or can talk well, or how well he can discourse on religion, and if you are feeling reasonable and don't have the impulse to discount someone by saying, "He's a sociologist, of course he can't talk," then you are likely to say, perhaps, "I don't know, I haven't met him." You might add that Dr. Grateoxe, who has met him, reports that he is a delightful conversationalist. Not so with Washoe. If we want to deny or assert that she is talking we tend to *think* about it instead of going to take a look and have a chat, and Roger Fouts, who has met her and says that she is talking, is discounted in a way that Dr. Grateoxe isn't.

Which brings me to a parenthetical issue. Normally, our sense of whether or not someone knows something has partially to do with our sense of their interest in and love for their subject—which is part of intelligence and integrity both. We prefer to have a mechanic who loves cars working on our engines, and a doctor who loves medicine working on our bodies. If the doctor loves people, too, so much the better. We prefer to learn philosophy from someone who loves philosophy. Love is not blind. But the animal trainer may be told that, because s/he hangs around the animals so much, the infection of sentimentality has set in, with the implication that familiarity and love breed ignorance. This is a difficulty we all face from time to time, and we may in fact invest ourselves in our subjects in ways that can lead to certain sorts of errors. Nonetheless, we trust the CPA who loves figures more than the one who hates them, while the trainer's love is occasion to doubt his or her account of what's going on. The burden of this creates in trainers a particular sort of soul-muddle, which is a kind of insanity. This is not directly my subject at the moment, but it is something anyone interested in this particular corner of the psycholinguistic show should be alert to—anyone, that is, with a sincere and civilized interest in finding out what the people who work with the chimpanzees and other apes in language research actually know.

The conversation with Washoe and Moja is about breakfast: "Do you want an apple?" "Give Moja fruit juice!" and so forth. I can't read Ameslan, or not much of it, but I experience, as do most people who happen on these conversations, a shock of recognition. This is language, I think, or at least what I call language. The pattern and immediacy of response seem unmistakable. I find that trying to have recourse to the "Clever Hans fallacy" as an explanation seems alien to my intuitive reading as a trainer.

But I am appalled and grieved because the chimps are in cages. This offends something. (And my project, which was to see with an ignorant eye, has failed. My opinions intervene, and I am miserable as a consequence.) What is offended is the dog trainer's assumption that language or something like vocabulary gives mutual autonomy and trust. I grieve, but not for Washoe behind her bars. It is language I grieve for.

Later I hear from Ken DeCroo, a linguist turned wild-animal trainer who has worked extensively with Washoe, the story of how Moja came to bite one handler's kneecap seriously. I learn from the account that when something unusual happens, chimpanzees, like people, feel an anxious impulse to *do* something, and that attacking the handler is an option that may readily recommend itself. This is not the sort of story I am accustomed to. Duke and Lassie may start out wild and uncontrollable, but they end up in the living room as respectable citizens. (This sort of story may offend someone who is moved by *Born Free*.)

Roger Fouts tells me at one point about Washoe's habit, when she was younger and less dangerous, of sitting in a tree in the mornings looking at *Playboy* magazine. (Apparently chimps have such tastes, though I don't know who encourages them; Lucy, in Maurice Temerlin's account, used *Playgirl* to masturbate with. I find this to be the most impressive evidence of all of the complex intelligence of chimps, requiring as it does quite a capacity for responding to approximations and representations.) There was a Famous British Philosopher visiting that year at the university. His route to campus took him past Washoe's tree in the mornings. And, in Roger's story, his philosophy broke down in the face of this compelling cynosure. I can see this easily enough. My own philosophy seems to be in danger of radical revision. But I don't know much about the revised philosopher, exactly how he was revised and whether or not the revision lasted. The stories are generally interrupted and incomplete. And I don't know how Washoe was revised by *Playboy*.

What has my attention is the cage, and the story about the broken kneecap. Stanley Cavell has pointed out that we don't have to talk to everyone about everything, but there are some things we do have to talk to everyone about if we are to talk to them at all. We have to talk to dogs about biting if we are to talk to them at all.

In Washoe's case, I find that I disagree with anyone who wants to say that because we can't talk with her about politics and art, it follows that what she does isn't language. We don't talk with four-year-olds about these things, either, yet we can place what they say in a continuum that includes political discourse. I can't talk with most of my writing students about the issues that face me these days in relationship to writing, and some of them may never have the requisite conceptual apparatus; this is not a reason to deny that they are writing. Nevertheless, we do have to talk to toddlers about attacking their playmates when that comes up, and I must, in order to work with a companion dog, be able to assume that he understands perfectly well the moral significance of peeing on the couch or of biting certain objectionable visitors. That is to say, under most circumstances he ought not to, even though he might want to.

Washoe, like my dog, has been told, and in no uncertain terms, that she ought not to bite even though she might want to. With my dog, the issue was settled long ago, almost without our noticing it, and we are in agreement. If my dog were to bite a visitor, I would be forced to consider the possibility either that the visitor had committed a crime or that my dog had gone crazy. And I would have to work out what had happened before I could again take my dog for a walk. If there was no reason for the bite, nothing that a reasonable person could recognize as a reason, the relationship with the dog would have broken down.

But there is no such agreement with Washoe, and Ken and Roger are, for the moment at least, still in some relationship with her. Ken tells me another story about Washoe attacking him. On this occasion she was charging for him. Ken, instead of defending himself or trying to correct her, signed, "Hug, hug!" Washoe, in Ken's account, hesitated in her charge and then continued forward—but forward into Ken's arms for a hug. I am reminded that Ken knows Washoe, and I don't.

Still, Washoe is behind bars. I don't know the end of the story, only that I am uneasy because it plainly isn't going to end the way *Lassie Come Home* does or *Our Mutual Friend*. But I notice at this point that my interest in Ken DeCroo and Roger Fouts is based in part on our mutual refusal to look to the animal behavior laboratory as it currently exists for enlightenment. Roger tells stories of meaningless horrors and degradations in the labs.

As well he might. I know a story that makes it clear that the animal laboratory is not going to produce tropes of community and communication. At my university there was an attempt to pass campus laws that would allow trained Companion Dogs to accompany their owners to classes, offices and libraries. This was a response to the rising crime rate on campus and to masses of evidence that indicated that dogs tended to put malefactors in the wrong mood and were thus a means of averting rather than encountering trouble. One of the curious things discovered in this situation was that blind students were not allowed to bring their Guide Dogs into science buildings. Because, shouted a choleric biologist, there are laws stating that any animal that enters a research building may not leave it alive! I don't know about that; what I am interested in is the biologist's astonishing, righteous anger.

There are probably genuine students of animals and communication in the laboratories—but how do you enter into a contract about talking with a being you are going to kill? (In the biologist's rhetoric there were, incidentally, richly elaborated tropes of the particular insanity, wildness and filth of animals—he was talking about Guide Dogs!—and this is, of course, neither my story nor Roger's and Ken's. But it is well to note that it is lively enough in more or less reputable corners of science and the law.)

I come, through listening and watching, to piece together a story about Washoe. It is the story the appalled dog trainer tells. I find parallels in Stanley Cavell's vision of Shakespearean tragedy rather than in cheerful tales about returning animals to the joyous freedom of the wilderness.

The chimpanzee trainer, or teacher, takes up with the young chimp. S/he works intimately with her, nurturing her, playing with her and teaching her to sign. Many wonderful things come of this, including a significant and powerful bond of love. The chimpanzee gets older, becomes sexually mature. If the chimp is Lucy, who lived with a family, then more and more limitations have to be placed on her life and on that of the family, and while limitations are not in and of themselves regrettable, I suspect that fewer and fewer guests are charmed when they are bitten.

The trainer, or teacher, or stepparent, still talks about how much s/he loves the chimp—and s/he does. Dealing with the chimp becomes more troublesome, but in the evening, over a beer, the handler talks about loving chimpanzees, and it is plain that a listener who cares is confronted with something that ought to instruct us all about love, rage and language. Othello, proclaiming his love for Desdemona, is no more convincing in his nobility, intelligence and love.

At the end of *Othello*, the husband has killed the beloved. At the end of the chimp story, so far as I know it, the chimp is behind bars. I supposed, rather stupidly, that this *was* the end of the story, that the handlers would, perforce, accept and live with the limitations of the relationship as they had thus far and make what they could of it. In part because I wasn't fully taking into account the nature of a mature chimpanzee, I thought this was an inexact analogy to my own case, in which my having a full relationship with my dog entails my living with limitations, including the fact that the dog can't read or drive me to the doctor when I'm ill, generally accepting the fact that the relationship is not an incomplete version of something else. It is a complete dog-human relationship. Accepted as such, it provides us both with what it is supposed to provide us with and has integrity—it is not something I need to do anything about. The dog fits.

But Washoe doesn't fit. Roger Fouts is working on a research project that he hopes will culminate in turning Washoe loose in Africa, with a band of other signing chimps, where they can be studied in a wild situation. The hunch is that having language will enable the band to survive despite their having learned no wild chimpanzee social and survival skills. It may work, and the news will be out: language is adaptive.

I am mystified by this; I want to sputter something like "But I thought you loved her!" (And would therefore want to keep her around.) I feel foolish, as though I were one of the people the Koehlers call the "humaniacs" who weep when they see dogs being worked, and it is clear that I am in the wrong story. Ken DeCroo says to me one day, as we are both standing outside Washoe's cage, "Our commitment to Washoe is over." Washoe, for her part, is signing hopefully on the subject of being taken out of the cage for a walk.

This looks and sounds a lot like marriage and divorce in cases where divorce is a substitute for the murder at the end of a tragedy. Othello kills Desdemona when language fails to give him complete certainty of her fidelity, certainty, that is, of Othello's safety in the face of the fact that she exists independently of him. Washoe, I find myself believing, is no more ready for this divorce, no more eager for it, than Desdemona was. Of course, there's a difference between the surfaces of the two stories that is not superficial. Desdemona wasn't unfaithful except in the way we are all unfaithful to the exact being of the Other. Washoe, on the other hand, will certainly maim or kill someone if precautions aren't taken.

So Washoe isn't, after all, talking? Not doing what I call talking when I assume that if you'll talk to me you won't kill me? I watch, early one morning, while Ken and Roger take her out of her cage for a walk. This entails the use of leashes, a tiger hook and a cattle prod. I am instructed to watch from a distance and to be very still. Ken and Roger don't take her very far—she remains within sight from my seat on the grass. I'm impressed by the precautions and think about going for a walk with my dog or a friend, and for a moment wish that is what I was doing.

But when the three of them—four, actually, since Washoe's adopted son Loulis accompanies them—are far enough away so that the restraint devices aren't visible if I don't stare very hard, I am struck with how very much the whole procedure looks like going for a walk. And do I have anything else to go on, beyond this small thrill of recognition?

Roger and Washoe squat down together and sign, discussing something they have noticed that I can't see. And I think that if any of my claims that the police-dog tracker is a citizen are to be met with respect, if what I claim is to have any coherence at all, then I must acknowledge that Washoe and Roger are talking—are doing what I call talking. I haven't forgotten the tiger hook, the cattle prod, the broken kneecap and the plan to send Washoe to Africa. But I am back to the conviction that I am looking at some condition of language.

And I am back to the feeling I started with, that the issue of what Washoe is doing, what condition of language we are dealing with, is not an intellectual problem, a puzzle. If I acknowledge that Washoe is talking, then of course I have to notice profoundly that language does not prevent murder. If language does not prevent murder, and if it may in fact cause murder, then I am at a loss. For I have nothing, really, but talk to go on. If the gestures and interactions of various sorts that I observe really do add up to "going for a walk," and if Washoe is dangerous despite that, then I may be thrown into confusion, may suffer, as Othello did, from skeptical terror, and may want to deny Washoe's personhood and her language rather than acknowledge the limits of language—which can look like a terrifying procedure. In the same way I may want to find a certain kind of relief by saying that rapists or the assassins of Anwar Sadat are religious fanatics or are in some other way inhuman, not of that kind of being in which I participate.

In any event, it is clear that we cannot *prove* that Washoe is talking, any more than Othello could prove that Desdemona was telling the truth when she professed fidelity. Nor, no matter how we riddle, puzzle and tease, can we prove that she isn't talking, so it may be best to leave off devising yet more clever professions of skepticism about the matter, and consider instead what kind of story we are constructing, and what kinds of stories are possible. While we consider, Washoe changes from minute to minute and day to day, as we do. Roger and Ken can't prove, on a given day, that it is safe to take Washoe from her cage, but they can "read" her, using the same criteria that I use when I am deciding how much contact it is safe to make with the man approaching me. If Washoe is doing a lot of signing, is willing to talk, that is some sign of safety—one of the very best, even if it isn't a guarantee. Roger and Ken and other people who work with the big apes live boldly, trusting language, speaking up in the teeth of the evidence of, as it were, her teeth, knowing that such boldness must fail in the face of Washoe's incomplete assent to the terms of the discussion. This is what we all do. This is what the Camp David* accord was about, speaking in the face of the failure of language to prove—to probe—the humanity, or personhood, of the other.

*This chapter was originally written some years ago. I wish that there had been since then evidence that major American officials knew more than Washoe does about talking, but there hasn't been, hence my having to resort to somewhat dusty references to full-fledged statesmanship.

Skepticism about whether or not Washoe is talking is not based on reasonable or rational considerations, and it may be that no one has yet discovered what such considerations might be. (How could Sadat have proved, before going to Israel, that Begin could or would talk?) All we can do is take a look, abandoning the "cover story," as Cavell calls it, and hoping instead to come up with a fiction that would make sense of what we try to say, about her and to her, realizing that there may be no such fictions about fully wild animals, except the sort Jane Goodall tells, which are about their territories, not ours. Perhaps we may someday domesticate the chimpanzee as we have the dog, but at the moment we don't have the story that will enable Washoe to spend her old age in a chair by the fire. That doesn't mean she isn't talking.

3

How to Say "Fetch!"

If you do not teach me I shall not learn.   BECKETT

Terms that have histories cannot be defined.   NIETZSCHE

If we consider, as I have been doing, the size and kind of the social space created by the language shared by two or more creatures, and if we describe the integrity of a language as the physical, intellectual and spiritual distance talking enables the speakers of that language to travel together, then it looks very much as though the dog and the horse (who are neurologically simpler organisms than chimpanzees and whose linguistic codes certainly appear simpler) have a greater command of language than chimpanzees do. There is even a sense in which a well-trained dog or horse may be said to have a greater command of language than a human being whose code is infinitely more complex. The dog/dog trainer language is perhaps more primitive (in the sense Wittgenstein has in mind when he criticizes Augustine or invites us to consider primitive language games) than the chimpanzee's language, or the schizophrenic's, but I can go a lot farther with my dog than I can with a schizophrenic, or a Nazi, if only because my dog doesn't bore me.

Does this matter? Or is it just a sentimentalization of the enslavement of the domestic animal? Well, dog trainers and horse trainers insist that training—teaching animals the language games of retrieving, say, or haute école in dressage—results in ennoblement, in the development of the animal's character and in the development of both the animal's *and the handler's* sense of responsibility and honesty. This is either hopelessly corrupt, in a sense that *The Genealogy of Morals* might unfold, or else it can tell us something about not only what goes on in training but also what it might mean to respond fully as human beings to "character," "responsibility" and "honesty."

It is worthwhile, then, to describe part of one of the language games of training, namely retrieving, in the hope of discovering what sort of moral cosmos is revealed thereby. I am going to begin with William Koehler's work, not only because Koehler is one of the great trainers in a tradition whose *locus classicus* is Xenophon's *Cynegeticus*, but also because his method is the most formally complete one I know. That formal completeness, together with his muscularly insisted-on freedom to talk anthropomorphically, gives his theory—his theology, if you like—a clarity and comprehensiveness rarely found since the collapse of the world views that made obedience a part of human *virtu*. We may read Spenser with pleasure, but how do we imagine ourselves obeying anyone or anything the way his knights do?

Koehler holds, against the skepticism that in the last two centuries has become largely synonymous with philosophy, that getting absolute obedience from a dog—and he means absolute—confers nobility, character and dignity on the dog. Dignity? This is such a repugnant notion in a world still reeling from the shock of the Third Reich that his books and his methods have had to survive legal challenges from animal lovers confused about what cruelty is. Now, even though the courtroom battles have been largely won, a serious segment of the dog-oriented population can't say his name without attaching to it epithets like "devil" or "monster." Very few people who think it's right to train a dog at all are repulsed when they actually see a dog being worked in the way he teaches. It is not his techniques but the morally loaded language in which they are embedded that repels his critics.

So, in order to consider the counters, gestures or utterances of retrieving language I am forced to consider a seemingly tangential issue, a larger one: What could possibly give us the right to say, "Joe, Sit!"—ever? In thinking about this there is a grave danger of wittingly or unwittingly invoking some sort of calculus of suffering—of saying, for instance, that although obedience training is uncomfortable or painful, it certainly isn't as bad as the fate that awaits the vast majority of untrained dogs, under the wheels of cars, in the decompression chambers of the Humane Society or in laboratories. Circumstances may force us, rightly, to apply such a calculus, but that is an emergency procedure irrelevant to a fundamental discussion of rights, including the right to the pursuit of happiness. (I take *happiness* to have at least the range of significance Aristotle saw in it.)

What sense does it make to speak of a dog's having a right? It may help to begin by asking what sense Koehler makes of it, in what sense, that is, he uses the term "right." He says that the dog has the "right to the consequences of his actions." What does this mean? Or what does it mean to be *able* to mean such a remark?

To be able to mean the remark, to take the responsibility for meaning it, is to be committed to imagining the natures of the commitments involved, or at least to acknowledge the possibility of such imagining; otherwise the remark is as deeply incoherent as the double-bind tyrannies of much preaching, including academic preaching. I believe the remark is coherent and that we can see a little of what it might mean to mean it if we consider the job "consequences" is doing here for Koehler, and for any trainer who uses the conception well. The trainer has occasion to be aware, as few people are, that human authority is corrupt to the core, and that any trope of ascendancy—especially the trope of nobility—stinks of the immodest, the self-righteous and the sadistic. Yet the trainer must get on with training the dog. The dog is compellingly present.

In order to get on with it, Koehler makes a sharp distinction between correction and punishment, understanding that the taint of punishment may be irrevocably in any authority, and that genuine authority must do something about this. This is not a distinction between lenience and harshness; it is part of a distinction between kindness and cruelty, or perhaps between rightness and cruelty. A sharp, two-handed, decisive upward jerk on the training lead, performed as impersonally as possible, is a correction. Irritable, nagging, coaxing tugs and jerks are punishments, as beatings are. The self-esteem of the handler gets into them, with the result that, by obeying or failing to obey, the dog takes on responsibility for the handler's emotional well-being, as we can make children or spouses responsible for our souls. This is the sort of obedience Lear wanted from his daughters. With some dogs, managing to exact a pretense of such obedience is as dangerous as it was for Lear with Goneril and Regan.

Corrections, in Koehler's vision, are administered out of a deep respect for the dog's moral and intellectual capacities. Punishments on the other hand are part—and this matters tremendously—of the demeaning repertoire of so-called trainers who propose babbling at the dog as sweetly as possible. Cooing, "Oh my goodness, what a GOOOOD doggie!" as one training manual actually suggests, is, for Koehler, profoundly cruel, dishonest and dishonorable, the flip side of a beating. Even moderately self-respecting humans grab their hats when addressed in such a fashion. And although dogs are on the whole surprisingly tolerant of our specious doctrines, many of them will, in effect, grab their hats or else, like Cordelia, attempt through precisely administered bites to turn the rhetoric. It is usually a diet of syrup, bribery and choked rhetoric, rather than physical abuse, that creates character disorders such as viciousness and megalomania in a dog. Biting is a response to incoherent authority.

It is difficult to see this through the sprawled and tangled rhetoric of the informal training most pets perceive, but highly trained animals have sufficient control to make the point with sometimes telling cogency. I am thinking now of Hans, a Doberman, one of the most talented and competent dogs I have ever known. His response to the command "Fetch!" was so instantaneous, accurate and powerful that it sometimes seemed the air must ignite as he leaped forward from his handler's side.

Among his more spectacular performances was the Drop-on-Recall. In this exercise, as performed in competition, the handler tells the dog to stay and moves some thirty to fifty feet away, then turns and, facing the dog, commands, "Joe, Come!" When the dog has traveled some distance, usually about halfway, a drop command or signal is given, and the dog must drop to the ground and wait for a new recall command. With Hans it was generally necessary to say, "Hans, Come down!" in one breath, for by the time the handler had finished pronouncing "come," Hans was already halfway home. And it was risky to perform the exercise on blacktop, since Hans responded to the command by simply flattening out in midair and sliding, accepting like a base runner the ripping of skin and joints the game of being a great dog entailed.

Even insensitive and inexperienced observers were impressed by this dog. But unfortunately, it was possible to be moved by Hans without understanding that you might have to earn the *right* to say, "Hans, down!" A person called Uncle Albert accompanied Hans and his handler one day. Uncle Albert decided it would be nice to have Hans do his "tricks" for Uncle in the absence of his handler, who had gone off to fetch some training equipment. Uncle Albert held out a liver snap to Hans while saying, "Come to Uncle," or some such folderol. Hans looked at him for a moment in disbelief and then, with a stiffness expressive of deep disgust, got up and walked slowly away, thus disobeying both the Stay command his handler had given and Uncle Albert's phony recall command. His handler, returning in time to witness disobedience of a sort that hadn't been possible for years, refrained from correcting Hans. (The complications of *that* moment require separate consideration.)

Bill Koehler puts the bribers and the coaxers together under one heading—"humaniacs." (To be distinguished sharply from people who devote themselves to the prevention of genuine cruelty.) Koehler calls humaniacs

> "kindly" people, most of whom take after a "kindly" parent
> or an aunt "who had a dog that was almost human and
> understood every word that was said without being trained."
> ... They often operate individually but inflict their greatest

cruelties when amalgamated into societies. They easily recognize each other by their smiles, which are as dried syrup on yesterday's pancakes. Their most noticeable habits are wincing when dogs are effectively corrected and smiling approvingly when a dozen ineffective corrections seem only to fire a dog's maniacal attempts to hurl his anatomy within reach of another dog that could maim him in one brief skirmish. Their common calls are: "I couldn't do that—I couldn't do that," and "Oh myyyy—oh myyyy." They have no mating call. This is easily understood.

This Nietzschean-sounding anger may in fact have one of its sources in Nietzsche, via Jack London, but exactly how Koehler got so angry is not my present subject, but what he is angry about—what vision of dignity and significance is at stake—is.

Another trainer, Jack Kersley, in *Training the Retriever*, writes:

> When a dog is corrected, perfect justice may be administered. There is no question of making an example so that others may be deterred, nor as in human crime must the penalty be made so heavy that the public is protected. Least of all must punishment be administered to satisfy the sadistic irritation of the trainer.... In nature, there are no punishments, only consequences.

Here "nature," of course, means something like "paradise," a region of clarity in which language never refers beyond ourselves and our intentions. Something very like a myth or story of expulsion from such a paradise stands behind the trainer's attempt to make sense of a life in which we must say, "Joe, Fetch!" or at the least, "Joe, Sit!" I hear a story that goes like this: When God first created the Earth He gave Adam and Eve "dominion over the fish of the sea, and over the fowl of the air, and over every living thing that moveth upon earth." Adam gave names to the creatures, and they all responded to their names without objection, since in this dominion to command and to recognize were one action. There was no gap between the ability to command and the full acknowledgment of the personhood of the being so commanded. Nature came when called, and came the first time, too, without coaxing, nagging or tugging.

Then Adam and Eve themselves failed in obedience, and in this story to fail in obedience is to fail in authority. Most of animate creation, responding to this failure, turned pretty irrevocably from human command. The tiger, the wolf and the field mouse as well as, of course, the grasshopper refuse to come when called, to recognize our naming. One may say that before the Fall, all animals were domestic, that nature was domestic. After the Fall, wildness was possible, and most creatures chose it, but a few did not. The dog, the horse, the burro, the elephant, the ox and a few others agreed to go along with humanity anyway, thus giving us a kind of second chance to repair our damaged authority, to do something about our incoherence. Training, in this story, can, through its taut catharses, cleanse our authority, for varying stretches of time, of Nietzschean *ressentiment*. Without that catharsis, dogs very properly withhold full obedience. Hence Hans's disgust with Uncle Albert, hence "wild and uncontrollable" Collies, and hence, of course, Fido's being perfectly happy to *play* fetch while refusing to bring the stick all the way back. The gap the dog insists on between us and the stick represents the gap between our ability to command, give advice and so forth and our ability to acknowledge the being of others. This is the taint in our authority. And the dog or horse trainer's special interest in "wild and uncontrollable" animals assumes that the degree of violence in a dog's resistance to authority may indicate the depth and power of response such a dog might give when the gap between command and obedience is closed. There is the mythology of Buck in *The Call of the Wild*, or of the dog Lion in Faulkner's "The Bear."

It is the full acknowledgment of language that closes the gap. Except for complete isolation, only such acknowledgment can deprive authority of its power to render false and sadistic the operations of our relationships. It won't do to suggest that the dog can just live peacefully around the house while we refrain from giving any commands that might deprive him of his "freedom," for that simply doesn't happen. We are in charge already, like it or not, and when a dog is about, everyone involved is going to be passing out advice and giving commands whether or not they've earned the right to do so. One might as well suggest that we leave off keeping toddlers out of the street or teaching anyone anything at all. We do assume authority over each other constantly, or at least we had better do so if only to be able to say, "Duck!" at the right moment. If our authority is weak, if we haven't taken responsibility for it, we *won't* say, "Duck!" at the right moment, or the person so addressed may not duck. A refusal to give commands or to notice that commands are being given is often a refusal to acknowledge a relationship, just as is a refusal to obey. The nagging punishments that require the dog to "want to please," to be responsible for the handler's feelings, fail to get authority out of the relationship by refusing to take the dog's position into account, or to bear the consequences of speech. In saying this, I am echoing Stanley Cavell's discussion of "Knowledge and the Concept of Morality," and I'd like to summon Cavell's aid further to say that Koehler's rage against humaniacs is the rage of someone who, knowing what the full acknowledgment of a dog demands from a trainer, must hear the for him unbearable "tone of one speaking in the name of a position one does not occupy, confronting others in positions of which one will not imagine the acknowledgment."

A practical consequence of the trainer's knowledge of the significance of command is that trainers say things like "Hans, Down!" far less frequently than most people, or at least to fewer dogs, and that they dream of a world in which they could say "Down!" or "Fetch!" more often, a world in which dogs had civil rights as a matter of course. If it sounds queer to talk this way, to suggest a connection between commanding and granting rights, that is only because we have failed to be sufficiently anthropomorphic. We don't imagine we can grant civil rights to human beings without first assuming authority over them as teachers, parents and friends, but we have lately supposed, strangely, that rights can be granted to animals without our first occupying the ground of commitment that training them instances. The dog trainer isn't willing to occupy that

ground with every dog s/he meets, can't, in fact (just as there
are certain responsibilities it is logically impossible to assume
toward someone else's spouse). And with this goes a refusal to
pretend to occupy the ground, knowing that such a pretense is
a debasement and deception of the dog, as any fraudulent
behavior is.

It is important to say that the trainer I am talking about is,
like the poet or the philosopher, an invention of our minds. I
have met people who think and act in relation to animals in a
way that is very nearly as coherent as that of the trainer I am
here imagining, but not even Bill Koehler is that trainer. However, there are facts that can be known about this imagined
object, just as there are facts about the mathematician's omega
that can be known, although the animal trainer is in some ways
more difficult to know than the mathematician's omega, because
the trainer speaks. In order to investigate this trainer further, I
am going to tell a story in which I am the trainer and the dog
is a year-old Pointer bitch named Salty.

Salty is a bird dog, and a good one, which means that the
chance of her biting me is so remote that we are unlikely to
have to discuss it. Not that Salty is a "soft" dog. She is a hard-
core Pointer, birdy and geared to travel all out for miles, the
heiress of an unsurpassable tradition that includes Algonquin,
the hero of Dion Henderson's wonderful book *Algonquin: The
Story of a Great Dog*, who was "very gallant, sir. I think perhaps
he would pity his bracemates if he were not enough of a
nobleman to know that they would rather die than be pitied."
For a true Pointer admirer, the Pointer's fire, speed, precision
and power constitute the central good in the universe, the rest
of which is judged by its ability to honor and celebrate the
Pointer. "Mr. Washington felt that any bird you didn't hunt
exclusively with Pointers was unfit to associate with gentlefolk."

So Salty, in my story, has it in her to be staunch to wing
and shot, to hunt wholeheartedly where lesser dogs lose spirit,
to work with full courtliness and gallantry, to become so dedicated to the perfecting of finding and pointing birds that the
sort of hunting where the art is adulterated by actually shooting
the birds too much of the time, for merely practical reasons,

becomes uninteresting. Unfortunately, her wonderful qualities may look to the uninitiated remarkably like wildness and uncontrollability. She's a year old, and she hasn't even had any puppy conditioning; the only part of the noble rhetoric she has gotten down is the part about going hard and tirelessly. This she does everywhere, as indifferent to obstacles like the kitchen table or the bedside lamp as her forebears were to nearly unleapable gullies and impenetrable thickets. Her former owners had known only the sort of dog who is more accommodating of cooing and bribery. Salty was not so much intolerant of the cooing as she was indifferent to it—it had, literally, no meaning to her. She thought she was doing a fine thing when she went through the picture window to point some meadowlarks in the backyard, and the cries of "You bad dog, how could you do this to us, you're supposed to be our precious baby, look at my potted Chinese palm, all ruined!" deterred her no more than the gashes on her neck and shoulders. But she was tolerant of the fussing, even if she was unresponsive to it. Proof of her tolerance is in her not having bitten anyone. A German Shepherd or a Doberman, or any of the breeds for whom the forms of companionship can be as deeply visionary as the forms of hunting are for Salty, might very well have responded to this particular incoherence with some hostility.*

When Salty enters my life she's been talked at a lot, but she knows virtually nothing of talking with people who mean what they say, so I don't say much of anything to her at first. I take her out of her kennel and am silent except for a calm "Good morning, Salty." I put her on a fifteen-foot line, attached to a training collar, and I begin to go for a little walk. Salty stays at my side for about one and one-quarter seconds, which is

---

*It is impossible to say enough in defense of the deeply civilized hearts possessed by most members of the working breeds, especially Dobermans. The fear of Dobies is all too easy to understand in light of the bad press they've had, which hasn't so far been offset by tales as powerful as those of Rin Tin Tin. I shall content myself by saying that the bad press was boosted by Hitler's charge that they were of impure blood and therefore unstable in temperament, and that perpetuating the myth of the vicious Doberman amounts to perpetuating a Nazi line.

how long it takes her to spot something huntable—in this case, my motley-coated cat, Touchstone, who is idly watching the progress of the ten o'clock shade of the pepper tree toward his sunny spot, contemplating as is his wont the curious ways the shadows move round and round, forcing him to change napping places as the day wears on. Salty heads for Touchstone at proper field-trial speed. I say nothing to her—nothing at all. Nor do I tug suggestively on the line to remind her of me. Instead, I drop all fifteen feet of slack into the line and turn and run in the opposite direction, touchdown style. Salty hits the end of the line and travels, perforce, some distance in the direction I am going, tumbling end over end.

This thoughtless behavior elicits no hysterics from her, as it might from a lesser dog who is used, as Salty is, to being coaxed and tugged away from kitty cats, instead of having to deal with a handler so clumsy and impolite as to fail to check first on what Salty is doing before dashing off as I did. She is startled, of course, but she forgives me and romps toward me to let me know it. On her way to forgiving me she catches sight of a leaf moving in a way that might be construed to mean *bird* and spins out after it. Again I head swiftly in a direction opposite to hers, and again she tumbles end over end, correcting her story about who I am.

The third or fourth time she gets dumped in this way, it dawns on Salty that there is a consistency in my inconsiderate and apparently heedless plunges. She sits down in order to think this over, cocking her head in puzzlement, trying to work out the implications of my behavior. She suspects, correctly, that they are cosmic. I respond by heading purposefully toward an open gate. She decides she'd better keep an eye on me and follows. Then the sight of the gate standing there in such an *opened* way tempts her and, forgetting the new cosmology, she charges for it. I turn again to do my routine, but this time she remembers before she's in high gear that I can't be counted on to follow, and she brakes and turns, loping in my direction and avoiding hitting the end of the line.

By the end of the first session she is more attentive to me, more willing to follow—to try, that is, to understand—than she has ever been to anyone in the whole course of her life, and I've said *nothing* to her. When the session is over, I utter the first command—"Salty, Okay!"—as I release her to think things over. The "okay" command is essential to establishing all sorts of clarities. It means something like "You are free to walk about, sniff things, take a nap, have some lunch, initiate a conversation with Touchstone on the mysteries of time, but you are *not* free to commit any crimes, dig holes, chase Touchstone, leap into the middle of my morning tea or generally buzz around at low altitudes." If she does any of these things I correct her by jerking sharply on the line while hollering "*OUT!*" in my most terrifying voice, sounding as much as I can like a clap of doom. Since she is neither a wild animal nor a hardened criminal but a true dog, she is able to grant fairly quickly the reasonableness of my demands, and since neither "Okay!" nor "Out!" requires her to love me or nourish my soul, her willingness to cooperate need not be contaminated by questions about whether she loves me better than she loves the vision of herself as a mighty huntress. There can't in any case be much talk of love between us at this point, though we do find each other likable enough. Love will follow on shared commitments and collaboration, on a mutual autonomy that is not now possible.

The mastery of the "okay" command is not an achievement of love but rather of the simultaneous granting and earning of some rudimentary rights—in particular, Salty's right to the freedom of the house, which, like my right to the freedom of the house, is contingent on making a limited number and kind of messes, respecting other people's privacy, refraining from leaping uninvited onto furniture and laps and making the right distinctions between mine and thine, especially in the matter of food dishes. I have managed to grant this right only by becoming inflexibly in command about certain things. It is a matter of authority, not love—I don't even have to like a dog in order to accomplish this. In most adult human relationships we don't have to do quite so much correcting in order to grant each other house privileges, but that doesn't mean that house privileges don't depend on the possibility of such corrections. Try putting your ice cream cone on my typewriter, and you'll get the idea. Freedom is being on an "okay" command.

Love, of course, is getting into things, but at this point I love not Salty but the literary tradition that has produced her. She doesn't know this tradition, but she has it in her to respond to and enlarge on my response to it in the way that a talented young poet has it in him or her to respond to a mentor's response. That is an important part of a talent for anything. In Salty I see echoes of the great field-trial dogs of the past and the paintings and tales of them, in the way I see in a student's work and manner of thinking echoes of the authentic energy of Stevens or Dickens. The better I am at responding with great accuracy—at, for instance, not making mistakes about which great tale or painting is being invoked—the greater the likelihood that this particular dog's qualities will be realized.

After about a week of work on the longe line ("longe" is a term for a certain piece of equipment, a long line, and for certain sets of exercises in both the dog and the horse worlds) it is time to introduce a piece of linguistic behavior whose syntax is more like the usual syntax of the imperative mode than "Salty, Okay!" is. I spend a while in an activity called "Teaching the dog what 'Sit' means," although that phrase captures very little of what teaching anyone anything is like. Briefly, I gently place her in a sitting position at the same time that I say, "Salty, Sit!" This is not what "Sit" will come to mean, but an incomplete rehearsal of it. I do this for several days, without giving any corrections, not because a dog needs more than one rehearsal to learn that the utterance means a certain posture and a certain relationship to my position, but because it *is* a rehearsal, and more than one rehearsal of more than one kind must precede genuine performance. There are, as I said, no corrections. Attending to me, refraining from biting and leaping about, is the only obedience required at this point.

Then comes the moment when I give the command without placing her (and without stooping or wiggling my ears or any other part of my anatomy pleadingly or suggestively). If she refuses to sit, or just doesn't think to sit, I give her a harsh, emphatic sit correction. The meaning of sit has now changed, been projected into a new context. Salty says to herself, "By God, I'd better sit when she says to, or the world will keep coming to an end!" The imperative mode has at *this* point, *this* form.

Someone may be wondering about all of that glorious dashing about, full expression of creaturely energy and so forth. It may seem that I've destroyed that, broken Salty's spirit. It is in fact possible to do so, although quite difficult in this particular case. With some dogs, syrup and cooing will do it, especially if there's also a full-blown emotional blackmail operation. That didn't work on Salty, but if I fail now not only in respect for her but in love of the literary tradition that has taught me to want to train Salty, I can in fact "break her spirit." Failing to give the corrections harshly enough would be an example of such a failure. When there is danger here, it is that so many of the stories we know about authority are about people getting stuck, as it were, in a parental or imperative mode. So with Salty I must be prepared to drop that instantly and to respond to every true motion with awe and recognition. If my "Way to go, kid!" or "Good dog!" invokes for me what I know and feel about the great dogs of the past, then what we are doing will lead in time to the second inheritance of her running gear. Dog training is one of the arts concerned with the imitation of nature, which is to say, the second inheritance of nature.

At this point in Salty's training it is not clear whether the utterance "Salty, Sit!" is language, even though there is plainly a looped thought involved: Salty, that is, is sitting in response to her recognition of my intention that she do so. But it's not clear how interesting or useful it is to *say* that it's language, or that anything much would be known thereby that we couldn't know just as well if the anecdote avoided terms like "meaning" and "intention" and used instead a Skinnerian or a Cartesian vocabulary. The situation is far more complex than my account of it, but at this point those complexities don't illuminate the issue. We have a looped thought, but the flow of intention is, as it were, one way. In my account the dog doesn't initiate anything yet. She obeys me, but I don't obey her.

One day, though, and quite soon, I am wandering around the house and Salty gets my attention by sitting spontaneously in just the unmistakably symmetrical, clean-edged way of formal work. If I'm on the ball, if I respect her personhood at this point, I'll respond. Her sitting may have a number of meanings. "Please stop daydreaming and feed me!" (Perhaps she sits next to the Eukanuba or her food dish.) Or it may mean, "Look, I can explain about the garbage can, it isn't the way it looks." In any case, if I respond, the flow of intention is now two-way, and the meaning of "Sit" has changed yet again. This time it is Salty who has enlarged the context, the arena of its use, by means of what we might as well go ahead and call the trope of projection. Salty and I are, for the moment at least, obedient to each other and to language.

Understanding has been enlarged by enforced obedience! How queer, one may want to say; surely this is a borderline case having very little to do with understanding as we usually talk about it. But our concept of obedience is larger than we admit in certain discussions when we're talking about the importance of autonomy or the power of the individual imagination. "Obey" itself comes from a word meaning "to hear." We covertly recognize that this may not be an irrelevant etymology by using expressions like "I don't follow you" or "I'll go along with that." It is a certain spookiness about the *word*, together perhaps with the terror of being suggestible and therefore weak, that makes us forget that no one is more suggestible (more responsive to a text) than the great critic or philosopher, no one more obedient, more fully surrendering of the self, than the master of an art. *What* we surrender to matters, of course. Salty appears to have no choice about surrendering to me, but to say that as though it distinguished her from us is to overlook the degree of choice any of us has about surrendering to our native tongues. My training methods, like any teaching method, depend on her willingness to cooperate,* which is one of the

---

*Since most house cats would rather die than obey a direct order, some people have been led to elevate the cat in their estimation, as T. S. Eliot did: "A dog's an easy going lout / He'll answer any hail or shout." The belief in the

reasons the chimp trainers want to insist on the word "teaching" for what they do, and there are contexts where they are quite right. Salty's surrender, I would like to say, is not entirely helpless. She can't know ahead of time how fully I will follow through on the commitments commanding another being entails, or even what it might mean to know that, but she has (unlike the wolf) something like the rough-and-ready criteria we use frequently in judging when to surrender. She has a million tiny observations of my tone and manner to go on (when people are as good as most animals are at this, we say they have a "way with animals"), and her recognition of them is not entirely unlike the recognition that leads me to attend more closely to a book I've picked up idly. Of course, I can always decide I've made a mistake and put the book back down (or at least I like to think this, sometimes forgetting that there are tiny increments of loss in even so trivial a repudiation), but Salty, too, can always decide to bite me and split. I have known this to happen. She is free, or rather she is not free, in the way babies aren't free.

Now, though Salty is not yet the master of anything, and I am not yet her master, some more rights have been granted to her, or she has earned them, whichever you prefer. If my behavior has been just, then Salty has the right to a certain attentiveness or responsiveness. This is part, as are the corrections, of what Bill Koehler has in mind when he says the dog has the "right to the consequences of his actions." We stand on the threshold of a discipline that can free us from some ancient and troublesome trespasses against language, and the resources of consciousness have been renewed.

There are a lot of things we don't have yet. Among others, we don't have a subject—a grammatical object, that is—for our

---

independence of cats is an example of what has been called anthropomorphism but is not. It is attributing to animals, not traits that we possess, but traits that we wish we possessed, or are afraid that we possess or that someone possesses. Pure savagery, unending and impersonal sexuality, unshakable love, etc. In this case, the mythical emotional independence of the cat. It is imaginative failure that obscures our view of the ways in which the house cat cooperates deeply in the domestic enterprise.

primitive language to engage. Even Salty's creative management of the food dish and the garbage can doesn't enable her to name the garbage can. Naming is an advanced activity of language and not the prior, essential act some of our allegories about ourselves and matters such as signing chimpanzees make it out to be. Names for anyone or anything but the speakers of a language are not necessary for knowledge or acknowledgment until we actually do name objects, and then they will be necessary.

We can now say something about how the story the behaviorist brings into the laboratory affects not only his or her interpretation of what goes on but also what actually does go on. To the extent that the behaviorist manages to deny any belief in the dog's potential for believing, intending, meaning, etc., there will be no flow of intention, meaning, believing, hoping going on. The dog may try to respond to the behaviorist, but the behaviorist won't respond to the dog's response; there will be between them little or no space for the varied flexions of looped thoughts. The behaviorist's dog will not only seem stupid, she will be stupid. If we follow Wittgenstein in assuming the importance of assessing the public nature of language, then we don't need to lock a baby up and feed it by machine in order to discover that conceptualization is pretty much a function of relationships and acknowledgment, a public affair. It takes two to conceive.

It turns out that saying "Sit" requires a lot of detailed work on various cases, for Salty's idea of its conceptual boundaries is as annoyingly rigid as the philosopher's idea of the conceptual boundaries of truth. She'll sit in the middle of the yard, but she won't sit a foot away from the fence. Sitting-on-grass is one thing, sitting-on-blacktop quite another. Sitting when she's calm is not sitting when another dog is inviting her to play, or sitting when the Mailman Monster is approaching, or sitting when she'd rather eat. I make her sit in as many situations as ingenuity and luck will provide, with some exceptions. For most dogs, sitting in a puddle of water is at this stage of training deeply unacceptable. There just isn't yet in the language available to us any way to make that a coherent demand. It can become coherent if we develop a story about hunting birds or trailing lost children to the kidnapper's den, a story whose domain includes the possibility of the articulation of a reason. (When is it coherent to give the command "Read Aristotle's *Metaphysics* by Wednesday"?—or "Take up battle stations"?) A great variety of projections of "Sit" are going to be required, and these are by and large projections into new moral contexts.

Similar projections of the meanings of terms will give the other novice commands a wide variety of uses. They are all unique in significance, but I want to look at only part of one of them, the command "Stand." In this exercise the dog learns to stand and stay while a complete stranger examines her (painlessly). This is the only exercise in which the dog must allow someone besides me to touch her, and the only one in which I must commit myself to staying within range for the duration—the other stay exercises are eventually performed with me out of sight, and I will eventually have her perform formal recalls from positions out of sight. In the out-of-sight exercises, the dog makes her own decisions about her safety, not being corrected, for example, for moving from a stay because someone touched her. But in the stand-for-examination, she gives over to me all of the responsibility for judging the danger of the situation. This is an extremely important exercise for, say, dogs in police work, because it greatly increases the dog's reflectiveness in judging situations.

At first I must be able to judge character as minutely as the dog does—and good dogs are by definition good judges of character. In particular I must be careful not to ask anyone who is a "natural bitee" to approach and touch her. Natural bitees are people whose approaches to dogs (and perhaps to people as well) are contaminated by epistemology. They attempt to *infer* whether or not the dog will bite, jump up on them or whatever. Instead of "reading" the dog, as handlers say (the German philosopher Martin Heidegger might call this listening to the dog's being), they cast about for some premise from which they can draw an inference that will give them certainty about the dog's behavior. They are—sometimes only momentarily—incapable of beholding a dog. It is not that the required information will follow too slowly on their observations, but that they *never* come to have any knowledge *of* the dog, though they may come to have knowledge *that*—knowledge that the dog is an Italian Spumone or a Komondor, for example. And dogs read this with the same uneasiness we feel when we walk into a room and find that our spouse, or a friend, has plainly been sitting around inferring something about us—welcome has been withheld. This creates in dogs and people an answering skepticism, an answering terror. The dog starts casting about for premises, making inferences back, tries to reach certainty, fails to reach certainty and sometimes bites, just as we do. Most dogs, however, express this uneasiness by trying to reassure us with affection, and dogs are astoundingly good at demolishing skeptical terror, which is why they are so often effective in therapy even when the therapist doesn't suspect that episte-mology is the disease. Dogs are in general more skilled at belief than we are. In a culture that so commonly equates skepticism about others with dignity, that is, the maintenance of personal worth and professional and intellectual integrity, the dog seems, as T. S. Eliot has it, "frequently undignified" and "much inclined to play the clown." The dog's clowning breaks through our murderous skepticisms in something like the way Dickens dealt with the tyrannies of "reason" and "fact" that kept the poor poor and dehumanized, and the rich just dehumanized. And it is not an accident that a circus dog named Merrylegs is the character whose absence marks the absence of any heart and genuine thought—any moral center, that is—in *Hard Times*.

However, Salty's clowning will be more effective and sig-nificant when she asks herself when it is appropriate, so I teach her the stand-for-examination. Salty, as a Pointer, has (as is evident in the still, brown-gold gaze she turns on fools) a bit more contempt for epistemologists than, say, most Airedales do, but she has no hidden premises from which to draw logically sound bites, so she is likely to break her stand by offering to romp, and she will be corrected. This will not only make company manners possible, it will mean that if the chips are down I can expect her to be more rather than less effective at biting in my defense when there's some good reason to bite.

If it sounds surprising to say that learning to refrain from attacking or romping is part of courage, that may be because we tend to forget that it does take courage for dogs to attack, and that both submission and discrimination are aspects of valor in most of the stories we've told about the various forms of life in which human heroism is possible.

This wonderful ability to distinguish between good guys and bad guys is related to a faculty I mentioned earlier, and will discuss at greater length in Chapter 4, on tracking dogs. Dogs have a highly developed sense of what does and does not belong in the dog-human world. With this faculty goes a tendency toward sometimes rather tiresome opinions, whose degree and form of expression will vary from breed to breed as well as from individual to individual.

Salty isn't especially opinionated, at least about hearth matters (quest matters look different), and she expresses her sense of the forms for the sake of wit. It comes about, for example, that I am in the front yard with her one day, working on recalls on a fifty-foot light line. She is doing beautifully because the increasing distance she travels back to me has brought with it a transformation. Her earlier pleasure in running to greet me when not distracted by philosophical cats, together with her glee in the use of that transcendent running gear, has been transformed into a love of the exercise for its own sake. She now performs recalls with passionate energy and precision. She is a pretty sight indeed. My friend glimpses this from the study window and comes out to watch her work. Salty, the next time I call her, performs a swift and flawless recall, a straight sit and an unbidden but accurate finish—but to my *friend*, who is idling about smoking his pipe. This is an ancient form of animal humor. Horses do it too, and every animal who thinks of the joke imagines the witticism to be original, but there are some dogs who wouldn't at this fairly late stage debase what are to them sacred forms, especially some members of the working breeds, who take the domestic virtues more solemnly. I knew, for instance, a German Shepherd who grumbled noisily at the other dogs when they didn't obey—a female, as it happens. My Airedale was another matter. When, in training him to retrieve, I began placing the dumbbell in odd ways, he enjoyed the play on form. At one point, for instance, I set the dumbbell on its end for the fun of it, and he scooped it up happily, tossing it in the air a few times on his way back with it, to show his appreciation of the joke. A very fine, very serious German Shepherd I worked with, confronted with the same situation, glared disapprovingly at the dumbbell and at me, then pushed it carefully back into its proper position before picking it up and returning with it, rather sullenly. The Airedale's enjoyment of oddity and ingenuity made him a good candidate for movie work, whereas the Shepherd is better suited for something like tracking.

All of this, it seems to me, bears on the question of animal rights, on what we have to notice when we project the incredibly complex syntax of rights, duties and the like into our various ways of discoursing with and about animals. The trainer welcomes the attempt to make the notion of "speciesism"* coherent but wants impatiently to say something like "Look! When the logician has said, 'for all x such that P, where P is the ability to feel physical pain, etc., etc.,' and goes on from there to propose a theory of animal rights and consciousness, s/he has said virtually nothing, has not taken the tiniest step toward imagining the personhood of any animal." I don't mean that there isn't something deeply appealing in the Australian philosophers confronting the scientists about what goes on with animals in the laboratories, as there is something appealing about arguing that whipping slaves is wrong, or that women ought not to suffer. Without opposing such activities, I want to digress for a moment to take a look at a small corner of the sort of thing that springs into a trainer's imagination when "speciesism" talk gets going.

Most people are told at some point that when confronting a strange dog the thing to do is to hold out your hand invitingly for the dog to sniff and not to pet the dog until you are sure

---

*"Speciesism" is a word some philosophers have coined as part of the claim that the logic of our oppressions and cruelties to animals is like the logic of racism or sexism.

the dog isn't going to bite you. This sounds innocent enough, especially when we first hear it, generally as toddlers. But it can be dangerous advice, since quite a few dogs may interpret the motion of the arm as a threat. (They are less likely to do this if the proffered arm actually does belong to a toddler.) Beyond that, it amounts to a discounting of the dog's creaturely dignity that is flabbergasting for a trainer. We have recently come to recognize that the assumption that you have the right to touch and speak to any member of a certain class of human beings is part of an oppressive rhetoric. And if I were to offer to shake hands and speak with anyone who came within reach I would be regarded as insane.

Most dogs—not all—are more forgiving of this than most people are, partly because they tend to take a lightheartedly opportunistic view of us anyway, and partly for something like the reasons women may go along with a story about women not having strength of mind, or whatever. It relieves you of the responsibility for your own soul, or seems to, and more to the point, it's usually rather difficult to figure out how the new story would go. When some "kindly" person interrupts a dog I am working with by petting and yammering, I feel a jolt of rage on the dog's behalf of the sort you might feel if a stranger or even a friend were to interrupt your concentration in the library or under the hood of your Spitfire and start telling you how pretty, or handsome, or whatever, you are. And in some moods I want to say that the failure to imagine why there might be such rage, what is really at stake, is one way of characterizing my sense of the failure of the philosophy of animal consciousness.

Most people are, of course, as innocent of the intention to, say, violate the animal's privacy or deny the animal's right to meaningful work as they are of any knowledge of the principles of animal training. In practice, however, trainers have come up with some unfriendly responses, such as "Watch out, he's had attack training!" or "Don't do that, he's *Not Your Dog!*" Educating-the-public is one pious possibility, but if one goes about all day expounding the principles of animal training, one gets no training done. Besides, there aren't any principles of animal training, only some aphorisms, dog stories and what not, just as there don't seem to be, if one looks closely, any principles of philosophy, just some insightful epigrams and philosopher stories.

Trainers distinguish between hard and soft dogs. Size and strength have very little to do with this; these are qualities of the soul. I have seen quarterbacks reduced to tears by Pomeranians. A hard dog is one who will give you a proper time of it in training, testing the coherence of your right to command at every turn. This hardness is very different from the wolf's resistance, which springs from the general inappropriateness of training a wolf in the first place. (A wolf is unlikely ever to use the sit exercise in the way Salty already has, expanding the possibilities of discourse between us, and this is hardly because wolves are stupid. Thus, the possibility of granting rights to a wolf is highly attenuated, but the wolf isn't especially interested anyway.) Salty's hardness springs rather from the courage to insist that it really matters how we talk. It looks hostile in places, just as Wittgenstein, writing out of stunning courage, looks a bit hostile in places. It is the hardness we see in the lives of people for whom a certain sort of coherence matters more than the conventions of approval. Someone might want to say that people of that sort risk madness or the exile of criminal status. So do hard dogs who don't find the right person to talk to.

From what we've seen of obedience so far, to say that it can be understood as a way of building a society of dogs and humans doesn't capture my sense of the difference between saying, "Joe, Sit!" and "Joe, Fetch!" The difference isn't absolute, but it's profound, and I want to explain to myself why there are people who can learn to say, "Joe, Sit!" pretty coherently, but who never learn to say, "Joe, Fetch!"

It may help to summon a distinction Auden made in his essay "The Virgin and the Dynamo," between societies and communities. He says a society is comprised of a definite number of members

> united in a specific manner into a whole with a characteristic
> mode of behavior which is different from the modes of

behavior of its members in isolation. A society cannot come into being until its component members are present and properly related. . . . [A] society is a system which loves itself; to this self-love, the self-love of its members is totally subordinate. Of a society it may be said that it is more or less efficient in maintaining its existence.

A community, on the other hand, is united, not by a system of relations, but by a love of something other than itself:

In a community, all members are free and equal. If, out of a group of ten persons, nine prefer beef to mutton and one prefers mutton to beef, there is not a single community containing a dissident member; there are two communities, a large one and a small one. To achieve an actual existence it has to embody itself in a society or societies which can express the love which is its *raison d'être*. A community of music lovers, for example, cannot just sit around loving music like anything, but must form itself into societies like choirs, orchestras, string quartets, etc., and make music. . . . Of a community it may be said that its love is more or less good. . . . A perfect order, one in which the community united by the best love is embodied in the most self-sustaining society, could be described as science describes nature, in terms of laws-of, but the description would be irrelevant, the relevant description here being, "Here, love is the fulfilling of the law." . . . [T]he traditional term for this ideal order is paradise. In historical existence where no love is perfect, no society immortal, and no embodiment of the one in the other precise, the obligation to approximate to the ideal is felt as an imperative "Thou Shalt."

So, the imperative "Joe, Fetch!" commands the dog (and the handler), not as Newton's laws were understood to command the behavior of falling bodies, but as God's laws command some. "Fetch!" cannot be said meaningfully unless it is said with reverence. Its coherence requires that retrieving be sacred for both members of the community. But here is a paradox: the trainer must speak as if the sacred spoke through him or her, as though training were prophecy, even while knowing that that is mostly impossible, that the gap between the sacred and our knowledge of it is ineluctable. This recognition is part of the responsibility taken on, and so a trainer seldom says, "Fetch!" and often tells (true) stories about the dog's being the ultimate authority as to the rightness of our methods. And if retrieving training becomes profane for a trainer, then that is that, and retrieving training either stops or becomes incoherent. One may say that Germany fell in love with Hitler, and beyond that, terrifyingly, that Hitler fell in love with Germany, that he was sincere, that he was able to command as a master commands a student of painting. Hitler, however, seems to have been less sensitive than trainers are to the importance of realizing that the ability to exact obedience doesn't give you the right to do so—it is the willingness to obey that confers the right to command. I suspect that people with this understanding seldom end up in charge simply because with it goes an awareness of the immense imaginative burden of authority.

In the case of dictatorships, Auden might want to remind us that there is also this consideration: "Of a community it may be said that its love is more or less good." Perfect love doesn't exist; perhaps our sense of uneasiness in the presence of what we call fanaticism may be expressed, not only, as Wallace Stevens had it, by talking about the "logical lunatic," the "lunatic of one idea / In a world of ideas," but also by saying that fanatics don't seem to have noticed that the world really is fallen, and that acknowledgment of this is as essential to our lives as that acknowledgment of human separation is to the prevention of tragedies in human love. Political tragedy, perhaps, comes about through failing to acknowledge imperfections in our apprehension of the sacred, what Cavell calls "the separation from God." Perhaps, too, this is why we feel that teaching a child to play the piano or (as in *Nicholas Nickleby*) to speak French *for profane reasons* is such a repellent travesty. The resulting noise offends more than good taste.

It happens that wolves have more emotional and physical armor against incoherent teaching from humans than either dogs or children do, so that incoherent piano lessons are much more common than attempts to get wolves to fetch. This doesn't mean that the wolf is any better at preventing in himself the attendant soul-muddles, but just that the trainer, faced with the wolf's teeth and the tenacity of his opinions, has more trouble convincing herself that any real training is going on. It is possible to form some sort of society with a wolf, but forming a community with one is at the outer edge of the likely.

When I have taught Salty the novice obedience commands, we have formed a society. There is some community as well, for it is in the nature of things that the precision of formal work stirs the Primary Imagination. But a curious thing has happened along the way. Salty has taken to digging holes, partly in order to occupy herself in the yard while I am at my typewriter, but also because hole digging is sacred; in the activity the secret significance of everything reveals itself. Here, corrections won't work. I may yell, scream, deliver "Out!" corrections all I like, but these will have little or no effect. She learns to make sure I am distracted, perhaps by listening for sounds of the typewriter, and has her hole-digging fix as often as she can. Any corrections and punishments are just part of the fun, accepted as a dedicated athlete accepts aches and injuries. I don't mean she *likes* being walloped, but she is not deterred by walloping as she was deterred when the matter of puddling on the rug came up: puddling on the rug wasn't sacred.

So I submit myself to the holy discipline of hole digging. Dressed in gardening clothes, I go into the backyard and discover the Hole. I rejoice. I dance a jig around the Hole in celebration of the Mystery. I congratulate Salty on the Hole and, still dancing, get out a spade and shovel with a view to making this perfect thing even more perfect. Salty is delighted and helps me dig the Hole. We perfect its Form, making it diamond- or heart-shaped. I dance another jig when we're done, and, still dancing, I go get the garden hose and fill the Hole with water. Then, still rejoicing, I put Salty's head in the Hole. She emerges quite quickly (she's a very strong, agile dog), gasping in astonishment and outrage. I am surprised and say, "But I thought you loved hole digging!"

I do this every day for three weeks. If there is no new hole, I redig the old one. It is not long before Salty starts hanging back as far as she can get when I start humming my hole-digging hymn as I get out my overalls. Her face begins to express something like "Christ! She's crazy! Hole digging is not fun!" And she stops digging holes and devotes herself to preventing the very thought of holes from coming into my mind. *This has nothing to do with either punishment or authority, and if it is corrupted by either, then it becomes cruel.* I just am, or have become, the sort of animal who has this crazy, incurable response to the sight of a hole; the only way to handle it is to keep me away from holes. (The spouse of a dedicated sailor may feel similarly about bodies of water.) An important aspect of this hole-digging cure is that it *won't work* unless I really do get excited in just this way about holes. If I get the idea that this is a battle of ego and stamina I'm doing punishment, not dog training. (Dogs have been drowned by people who didn't *get* this.) Merely going through the motions won't compel the dog's belief. Holes must become sacred for me. This obviously means that unless some other object, such as tidy lawns or flower beds or the safety of guests, is stronger in my imagination than hole digging, I will lose my ability to disapprove of holes, perhaps permanently, and in the future chances are quite good that unless the foundations of the house are threatened I won't cure any more dogs of hole digging and will instead stop and admire their holes. I may think up techniques for training dogs to dig holes on command.

One day Salty and I are going for a walk through an abandoned orange grove. We come upon a hole made by some burrowing animal. Salty glances at me nervously, hoping I haven't noticed, and when she thinks I am distracted, reaches out a nervous paw and scoops some dirt and leaves over the hole. Then she frisks off, inviting me to play elsewhere. There is more than grammar in the symmetries and reciprocities of a training relationship.

Trainers tend to talk about the importance of corrections being impersonal, especially the out-corrections I discussed above. That's right, though the term is a bit misleading; it would also capture something to say that corrections should be as personal as possible, that they should be expressions, not of opinions, but of the trainer's nature. You simply become the sort of animal who, as it were, helplessly gives certain corrections in the face of certain crimes. This is something like the impersonality of the law, having to do with our sense that the law ought to be sacred to judges, but it also has to do with our sense that a good judge, or a good teacher, is not so much someone who is good at slipping into the imperative mode as someone who can do it without expecting that with obedience can or ought to come obeisance as well. Alas, the law is administered, and music appreciation is taught, by and to people for whom they are not sacred. There are ways of diminishing the resulting incoherence, but no way of eliminating it. Kurt Vonnegut pointed out the extraordinary power of denial the ghetto student acquired simply by thinking to say, "Whaffo ah wanna read no *Tale of Two Cities?*" It was not that there was no meaningful answer to this question but that most of us, living in the complacency Nietzsche tried to reveal for what it was, had forgotten that there could be such a question and that having an answer to it, or as many answers as there are people to whom one assigns reading lists, is part of the obligation taken on. Frequently the only answer that can be given ahead of time is: "Because you have to if you want to pass this course. You may also fall in love with Sydney Carton, or learn something useful about the necessity and tragedy of revolution, or you may come to love Dickens' language. Or none of this may happen." A teacher may or may not be willing to accept the further commitment of entering into the student's reality to see what meaning Dickens might have there. Not much else can be said, perhaps.

Therefore, when I go to teach Salty to retrieve, I don't speak to her of the greatness of Algonquin, although I need to remember it myself in order to respond appropriately to the minutest motions of such greatness in what goes on. They will be pretty minute for a while. First, I command her to sit and stay. I gently place a dumbbell in her mouth for a moment as I say, "Salty, Fetch!" I remove it instantly and praise her. She tightens her chops and regards me dubiously, perhaps swallowing to get the taste of the wood out of her mouth. *This* has nothing to do with the passion that led her to tear through brambles in order to chase balls for her former owners. At this point she can't even say, "Whaffo ah wanna fetch?" because the word doesn't mean that yet. All it is, is part of a distraction from the stay command.

For about a week I place the dumbbell in her mouth several dozen times a day, saying the command. Not because it takes her that long to "know" in some abstract sense *what* I mean, but because she has to know that I mean it. She can ask, "Whaffo ah wanna let you put that dumb thing in my mouth?" and I can say, "Because if you resist you will inevitably break your sit-stay and the earth will open up and swallow you." That's all we can say. She can't believe formal retrieving language at this point, in something like the way I can neither believe nor doubt Maxwell's equations. Anyway, even if I could explain things, the explanations would no more put truth into the commands if it wasn't already there than the word "true" will put truth into propositions if it isn't already there. Frege, the founder of modern mathematical logic, remarks in an awesomely lovely essay, "Thoughts," that "assertoric force does not lie in [the word 'true'] but in the assertoric sentence form; and where this form loses its assertoric force the word 'true' could not put it back again." This is perhaps truer of commands than of declarative sentences. I had better be right, that is.

There is now an object in our language, the dumbbell. After requiring Salty to allow me to open her mouth and place the dumbbell in it, I require her to open her mouth herself. Then I demand that she actually reach one full inch, and after a week six full inches, and so forth. At each stage I get, "Whaffo?" All I can say is, "Because I'll pinch your ear if you don't." My praise is another matter. When she does accept the dumbbell, I must respond with the awe that consists in honoring the details. Here the contaminations of approval (as opposed to recognition) will be worse in their effects on the language than with the novice exercises. Salty may be willing and able to sit in order to please me, but even if she wanted to she couldn't retrieve in order to please me. Even if I wanted to I couldn't write philosophy in order to please you. If I am writing philosophy I am presumably prepared, as Wittgenstein said I must be, to discover something completely new, and this completely new thing may not be—probably isn't—among the things that already please you. Hence no matter how large and wonderful and full of glorious creatures your philosophy already is, and no matter how much you love me and approve of me, and no matter how much this love and approval is welcomed and returned, it will be impossible to do philosophy in order to please you. I may, of course (will, in fact), be deeply helped in my writing by the conviction that you are able to recognize what I have done or am trying to do, or at least that you are able to acknowledge without demanding prior proof that I am trying to do something that will count as philosophy. This is the sort of praise I must be able to give Salty if we are ever to trope this or any other version of retrieving into a full-blown true story about her independence of mind and greatness of soul. Full-blown retrieving demands full-blown love of the activity itself. Salty doesn't (can't) retrieve *for* me, she can only retrieve *with* me. This is a mentor relationship. I am not denying that encouragement is essential to knowledge and achievement; we do take heart, are encouraged by each other; we learn things through each other by heart. However, the sort of encouragement that leads to a novice working with a nervous eye out for teacher's approval is a distraction from the business at hand. In practice it is probably not possible to be interested in what the novice is doing without being interested in the novice, or being willing to be interested if that becomes appropriate, but that is not to the point.

I should notice at this or at some point that one of the things that might lead someone to wonder about what looks like the wildest sort of anthropomorphizing is the sketchiness of the tokens of this language game. One thing I should say is, I'm not filling in all of the details (this isn't a language primer). More to the point, a reason for trying to get a feel for a dog-human language game is that it sharpens one's awareness of the sketchiness of the tokens of English. Wittgenstein says, "It is as if a snapshot of a scene had been taken, but only a few scattered details of it were to be seen: here a hand, there a bit of a face or a hat—the rest is dark. And now it is as if we knew quite certainly what the whole picture represented. As if I could read the darkness." When we learn a language game, we learn to read the darkness.

Retrieving makes possible a new sort of truth between Salty and me. It also makes new sorts of deception possible, and, in fact, the new truth is possible against the possibility of the new deceptions. The day comes when I lay the dumbbell on the ground, move about a leash length away from it, stand with Salty sitting at my side, the two of us facing the dumbbell, and, without my hand on the dumbbell, send my dog. She leaps out eagerly, with every appearance of sincerity, swiftly covers the dumbbell with her chest, lying on it, and then sniffs around in all directions industriously as if to say, "I'd love to bring it to you, boss, but I just can't find it!" It is unlikely that she doesn't know where it is, so I get on her ear and correct her, and she screeches with the sting and indignation of it. Suddenly she "remembers" that it's under her chest and picks it up.

This is a moment when both of the Koehlers would be especially disgusted by any hesitation to correct, and the word they use to express their disgust is "dishonest." Why? Salty's dishonesty is clear enough, but why would it be so wrong to fail to correct her? Why wrong to say something like "Well, the poor thing is just so upset by the newness of everything that I'll wait a while before insisting"?

By now I am saying a great deal when I say, "Salty, Fetch!" Not everything I will come to say, but a great deal, and one of the things I am already saying is "I promise that something is going on here that is worth doing right, and I am deeply committed to getting it right, and I know it is appropriate for you to pick up the dumbbell when I command you to." So in failing to correct I'm going back on my word, just as I would be doing if I corrected angrily, righteously or pettishly, as if I were personally offended. Similarly, if a student poet performs some anxious evasion of form and if I know (this is knowing a *lot*) that for this particular poet a full surrender to this particular form will be fruitful, and if I then accept the evasion, fail to correct, perhaps because I have a headache that is making it hard for me to concentrate or to care, then the teaching relationship will have moved toward incoherence and may disappear. This happens all the time, of course, and I seldom know enough about any student to be that certain of my corrections. The moral of this story, however, is not that we ought to lessen the demands we make on poetry itself.

I do know enough about dogs, and about this particular dog, to feel confident of my corrections and my manner of praise. Soon Salty explores another form of dishonesty, whose syntax is more like the syntax we have in mind when we think of lying as opposed to deception or evasion. Now I am throwing the dumbbell out as far as twenty or thirty feet, and it lands between Touchstone and a stick from the pepper tree. Salty picks up the stick rather than the dumbbell. This is almost like saying, "It's not this one but that one," although there are some queer things about it. For one thing, I assume that Salty doesn't imagine that *I* don't know the difference between the stick and the dumbbell. Salty is lying about *herself,* not about the dumbbell. She wants me to believe that *she* believes I meant the stick rather than the dumbbell. A curious thing about dogs is that the more talent they have for retrieving, the more they tend to think of inventive variations of this maneuver. Salty is very talented indeed, and she tries to retrieve Touchstone, rocks, the tires to my car, the lawn. She looks a bit sullen when I don't accept these "retrieves." It is as if the forms are so deeply if inarticulately felt for her that she must test my commitment to them thoroughly. That, at any rate, is a story that will enable me to get on with training. Another story is that she is playing around with the story I am trying to tell, is curious enough about it to be willing to risk the corrections.

What I want to do is note the nature of this "lying." Since she doesn't yet know the honesty of a proper retrieve, because she hasn't experienced it, she can't very well lie about that, and not much is being violated, by her at least, though a great deal is violated if I accept her lies as truths.

One day I notice that the nature of her retrieving has changed. I can tell, by the knowing way she sails out, the purposefulness of her movements, the wholly gay seriousness

with which she scoops up the dumbbell, the addition to her
performance of a degree of precision and fire I hadn't asked for
(since no one can ask for *this*), that it's Happened. She has
walked, or galloped, into real retrieving. She is transformed, I
am transformed and the world is transformed, for now I am
able to mean all of *this* when I say, "Salty, Fetch!" Now there
are all sorts of new ways our language can be projected. I have
her retrieve things besides her dumbbell. Perhaps I have her
retrieve Uncle Albert. Retrieving can become carrying messages. I can teach a directed retrieve, having her retrieve things
I haven't thrown. I can, that is, use "Fetch" to name things,
in somewhat the way we use "this" and "that" to name things.

If Salty and I proceed to scent work, then many other moral,
syntactical and theological complexities will enter the situation,
including a more advanced syhtax of deception and hence a
more advanced syntax of truth. Now, when sent to select one
of two or more indicated articles by scent, she can say, "It's
this one (rather than that one)." I don't see Salty as a cold
trailer, but she might be and might come to say, "The child
went this way," or "Here is the criminal's hideout." The investigation of how this comes about belongs elsewhere; I mention
it now in order to point to what I have been hoping to make
clear all along. The investigation of animal consciousness, like
the investigation of human consciousness, is centrally an investigation of language, and this ought to remind us of what an
investigation of language is.

The beginning of scent discrimination involves having the
dog look for hidden objects whose location the trainer knows.
So in scent work it becomes possible to give *advice*, and to give
advice about something we are mostly ignorant of (scent).
"Look for the criminal's tracks over there, to the west of the
sycamore tree," or "The way the wind is today, we'll do better
if we move downriver." This is what generally spoils scent
work. Once in a while the advice is right, but it almost never
is, and it almost always discounts the dog's possibilities as an
autonomous, trustworthy, responsible creature. Knowing nothing about scent, people give advice about it anyway, because,
I think, we tend to do this with each other, and it's just that

the "irrationality," as Cavell puts it, of the way we give advice—"As though advice operated on others at random, like a ray"—is generally more obvious in scent work.

How advice comes to be possible isn't my present concern. That advice is given brings me to notice that in the natural history of this language I am looking at, the possibility of giving a command is prior to (both historically and in terms of the unfolding structure of the syntax) such things as naming, referring, advising and full trust. (I now trust Salty enough so that although the general rule is no-dogs-on-the-furniture, I know that if she should once in a blue moon get up on my chair, she is making some sort of joke.) Trope is possible, too. (One day I absentmindedly fail to respond when Salty brings me her feed dish, a projection of "Fetch" she thought of. I put the dish on the desk and continue my work. Now she brings me a wastepaper basket, wriggling so hard with gleeful appreciation of her own wit she drops it on her way. I don't know what trope it is.) There is a great deal to say about this, but all I want to say here is that I am struck by a new wonder at the priority of commands and also at how the coherence of the commands depends on my ability, my willingness, to hand authority over to Salty, in the case of the wastebasket by acknowledging the possibility of her saying something I haven't taught her to say. (An extraordinary number of failures at formal tracking trials happen when the handler pulls the dog away from the trail, which may be one of the reasons the bigger, stronger dogs do so well at tracking.) If I can't say to you, should the occasion arise, "Duck!" or "Put the broom in the closet when you're done," or "No mayonnaise on mine," or "Lower down. Ah, that's it," then it's not clear what sort of relationship we can have. What if I can't give the command "Stop!" In dog training, commanding is made possible because dogs and people are domestic, and this is true without formal training, too, of course; there are all sorts of gestures and commands pet dogs respond to. Those are the commands being imitated in formal training, imitated in some of the ways that poetry imitates everyday utterance, which is part of why trainers have the impulse to speak of the *art* of training.

In cheerfully suggesting that authority is essential in our relationships, logically essential to speaking, that talking depends on the possibility of command, I haven't forgotten the taint in our authority I began by worrying about. If commanding is essential and if the untainted expression of authority is well-nigh impossible, since we can never fully know another's cares and interests, never fully share the sacred and profane objects in their world, then how *are* we to command each other? How say, "Turn in your papers on Tuesday," or "Take one pill after each meal," or "Practice an hour a day," or "Marry me!" or "File your taxes by April 15?" What gives us the right to say "Fetch!"? Something very like reverence, humility and obedience, of course. We can follow, understand, only things and people we can command, and we can command only whom and what we can follow.

4

Tracking Dogs, Sensitive Horses and the Traces of Speech

The horse, as it stands, is a rebuke to our unreadiness to be understood, our will to remain obscure.... And the more beautiful the horse's stance, the more painful the rebuke.

STANLEY CAVELL

With tracking and other kinds of scent work, dogs and people behave in ways that give fairly clear and obvious answers to one naive (philosophical) demand for a syntax of truth. In formal competition in this country, dogs entered in Utility Dog trials under the auspices of the American Kennel Club are sent by handlers to pick out from among a group of objects scattered on the ground the one bearing the handler's scent. The dog picks up one of the articles, thus saying, "It's this one (rather than any of the others)," and whether or not it *is* the right one is a function of whether or not the dog is lying. But that aspect of the syntax of scent work is not what is interesting and difficult for the trainer, to whom the dog's capacity for truth telling and lying is no news. At least, that is only the skeleton

of what is interesting about it, essential to it. As philosophers
entranced with the fact that creatures with language can utter
propositions that are either true or false under Convention T,
we may take the skeleton for the fundamental articulation, but
it is not the skeleton that supports the body of language, as
anyone who has ever tried to make a pile of bones stand on its
own ought to be able to guess.

The interesting and essential aspect of syntax that is revealed
by certain contemplations of tracking is a part that was never
concealed, or at least never absent. Almost anyone can indicate
our usual and strangely panicked awareness of this by remembering that we sometimes talk about how dogs can "smell
fear," as though we couldn't know fear in others, and as though,
further, the dog and our stories about dogs located the ways
in which cosmic objections to fear were punitively gathered.

Similarly with horses: They *know* when you're afraid! And
then there is all the advice about how you shouldn't approach
or try to deal with them if you are afraid. Because they will
KNOW. My husband and I, recently translated from California to the Northeast, were given similar advice about
deciding whether or not to walk through Spanish Harlem. If
you think it is dangerous, then take a bus or a taxi, because
if you think it's dangerous, then it is dangerous—they can
tell. And people present me with stories about horses who
are fine with their own riders but strangely difficult and/or
violent with less tactful or skilled riders, as though there were
some mystery to a horse's objecting to a stranger's taking
undue familiarities. The assumption that horses, dogs and the
inhabitants of whatever neighborhood is not your own are
possessed of powers of judgment whose principles are different from ours is part of a general tendency we have to suppose
that there are creatures somewhere out there who are relieved
of the limits of knowledge that we must deal with continually
and thus relieved of the burden of personhood, like the dog
in a horror movie who knows the horror before anyone else
can guess at it. Cavell points out that "it is important that
we do not regard the dog as honest; merely as without

decision in the matter."* Trainers, of course, do regard dogs and horses as either honest or not, as mean, sneaky, kind, sane or insane, and they know that they can make mistakes in a given judgment about an animal's honesty, so one way of understanding training is as a discipline in which one learns more and more about a certain steadiness of gaze, a willingness to keep looking, that dismantles the false figures, grammars, logic and syntax of Outsiderness, or Otherness, in order to build true ones. In the case of animals, I mean—for trainers, as for everyone else, humans (usually) are harder to work with, informally at least.

While dogs and horses do have perceptual capacities or kinds of awareness that we don't, the dream of the great tracking dog or of the transcendentally sensitive, alert and bold horse is not a dream of horror by and large, but something more like a dream in which the familiar and the beyond, the hearth and the quest, learn not confluence and identity, but respectful possibilities of welcome, even though that dream may be truncated into the foundation for a horror story, just as truncated and naive assumptions about marriage are. There is, nonetheless, a discoverable logic of welcome.

With almost any animal work, we begin by learning, gradually or suddenly, that for us seeing is believing. Aristotle opens the *Metaphysics* with a small warning about this, and the trainer, writing about or teaching tracking, warns that if you do not learn this, you do not begin. What skepticism largely broods about is whether or not we can believe our eyes. The other senses are mostly ancillary; we do not know how we might go about either doubting or believing our noses.

For dogs, scenting is believing. Dogs' noses are to ours as a map of the surface of our brains is to a map of the surface of an egg. A dog who did comparative psychology might easily worry about our consciousness or lack thereof, the way we worry about the consciousness of a squid.

*Stanley Cavell, *The Claim of Reason: Wittgenstein, Skepticism, Morality and Tragedy* (New York: Oxford University Press, 1979).

We can draw pictures of scent, but we don't have a language for doing it the other way about, don't have so much as a counter for a representation of something visual by means of (actual) scent. We cannot know, with our limited noses, what we can know about being deaf, blind, numb or paralyzed; we do not have words for what is absent. If we tried to coin words, we might come up with something—like "scent-blind." But what would it mean? It couldn't have the sort of meaning that "color-blind" and "tone-deaf" do. "Scent" for us can be only a theoretical, technical expression that we use because our grammar requires that we have a noun to go in the sentences we are prompted to utter about tracking. We don't have a "sense" of scent.

We do have some uses for the word in non-tracking situations, of course. We use it to mean perfume and such, and we use it as a figure of knowledge, as when in literary contexts a lover or an enemy speaks of someone's scent. But that is a metonymy for knowledge usually, the way the notion of a Bloodhound nose is used in detective fiction to refer to intuition. What we have is a sense of smell—for Thanksgiving dinner and skunks and a number of things we call chemicals. There is about as much relationship between our sense of smell and scent as there is between my ability to know about blasting through tremors in the ground and the knowledge of a great musician or a conductor—a knowledge that does not end if the musician becomes deaf. Or between a mole's sensitivity to bright lights and the seeing/thinking of Michelangelo.

So if you and I and Fido are sitting on the terrace, admiring the view, we inhabit worlds with radically different principles of phenomenology. Say that the wind is to our backs. Our world lies all before us, within a 180° angle. The dog's—well, we don't know, do we? He *sees* what we see—his eyes aren't defective—but what he *believes* are the scents of the garden behind us. He marks the path of our black-and-white cat as she moves among the roses in search of the bits of chicken sandwich I let fall as I walked from the house to our picnic spot. We can show *that* Fido is alert to the kitty, but not *how*, for our picture-making modes of thought interfere too easily with falsifyingly literal representations of the cat and the garden and their modes of being hidden from or revealed to us. I say "falsifyingly literal," not because the literal is automatically false, but because our attempts to think about scent locate one of the areas where the mind's lust for the literal, what Wittgenstein called "grammar" and Frege "speaking," misleads us in ways we can find out about. My very impulse to say, for example, that thinking about the problem of tracking "clarifies" or "brings into focus" certain aspects of language suggests how profoundly oriented I am to sight.

But here is the thing: We *do* talk with dogs about scent, and some people do it quite well. It is uncanny when you first point at some bit of emptiness on the ground and say, "Go find it!" It is possible to feel quite foolish and uncertain indeed, pointing at nothing as though you knew it were something, and avoiding that particular sensation of foolishness is a powerful impulse in human beings. L. Wilson Davis remarks in *Go Find!* that "even the most skillful professional handler will occasionally yield to the temptation of asserting his intellectual superiority over the dog. More often than not, it is the handler who defeats the dog, rather than the difficulty of the trail." Trainers disagree quite a lot about exactly what counts as handler error, but they all agree that supposing that you have to know what you're talking about when you say "Find it!" is a failure.

This is, like most failures, a failure to trust language itself. The handler has said, "Find it!" and the dog is trying to understand (follow, obey), but the "it" has nothing that the logician in us would recognize as a referent, so the woeful handler, unable to mean what she says without knowing what she reaches toward with her words, has failed to speak fully. The dog, more skillful at belief than the handler is, usually tries anyway, however briefly, but won't try forever.

Research on scent reminds us of how little our modes of thought can analytically probe the modes of thought that are occasioned by scent. Here is part of what happened when the U.S. Army developed an interest in trying to figure out how to defeat the enemy's trained tracking dogs:

In these tests trails were laid through "impossible" terrain...
large fields were sprayed with gasoline and burned after tracks
were laid through them. Foreign odors, organic and inorganic,
attractive and repelling, were introduced on top of previously
laid tracks. Tracklayers were picked up in cars and driven
several hundred yards where they continued the tracks, or
they removed their shoes along the trail, replaced them with
a sterilized pair, and continued the track. Tracklayers entered
rivers, swam downstream under water with snorkels and
emerged on the opposite banks to continue the tracks. None
of these tactics or devices consistently defeated the trained
tracking dogs used in these tests.

You can have ten people enter a room and mill about in it
for a while, then one of them leaves. When you bring your
tracking dog into the room and ask her to search, she may very
well nose about for a while, sorting things out, and then take
off after the missing person. There have been tests involving
skunk spray, solutions of alcohol, formaldehyde and such, in
which sebum and hair clippings were preserved for several years.
Trained dogs did not fail to match an object on which a few
drops of the solution were released with an article handled by
the person whose hair had been used in the first place.

But there is something to keep in mind here: I don't know,
in most of these cases, whether the handlers know the answer
to the problems being posed for the dogs; answers to questions
about scent have a logic that depends, as the logic of all questions
and answers depends, on the shared knowledge of dog and
handler. This is *not* to suggest that the so-called Clever Hans
fallacy is undermining all results, or even most results, in scent
work.* (I suspect that one of the implications of this might be
important to the problem of identity—having the right answer
to "Is this the same as the other thingummy we were consid-
ering?" depends in part on two individuals who speak the same
language, share the same form of life.)

Failures in tracking are often training and/or handling fail-

---

*In formal tracking trials, to cite no other instance, the handlers do not know
where the track is.

ures, which means that the marvel here isn't the dog's scenting powers, isn't, that is, a technological wonder but a moral one, as a small amount of anthropomorphizing ought to make obvious. The tracking dog must concentrate intently and ignore distractions, and Xenophon's catalogue of the faults of hounds shows that emotional distractions such as discouragement or loneliness or offers of sympathy are, for dogs as they are for people, on a par with distractions of the more enticing sort; his list sounds like a remarkably copious catalogue of human defects in character, with a few exceptions. For example, while impatience in various forms is a fault in a hound, that form of impatience whose opposite virtue in humans is a capacity for delayed gratification is not a vice. For dogs as for artists, in a roughly Romantic tradition, nowness is central most of the time.

Here are some of the faults and idiosyncrasies Xenophon cites:*

> Moreover, hounds of the same breed vary much in behaviour when tracking. Some go ahead as soon as they find the line without giving a sign, and there is nothing to show that they are on it. Some move the ears only, but keep the tail still; others keep the ears still and wag the tip of the tail. Others prick up the ears and run frowning along the track, dropping their tails and putting them between their legs. Many do none of these things, but rush about madly round the track, and when they happen upon it, stupidly trample out the traces, barking all the time. Others again, continually circling and straying, get ahead of the line when clean off it and pass the hare, and every time they run against the line, begin guessing, and if they catch sight of the hare, tremble and never go for her until they see her stir. Hounds that run forward and frequently examine the discoveries of others when they are casting about and pursuing have no confidence in themselves; while those that will not let their cleverer mates go forward, but fuss and keep them back, are confident to a fault. Others

*In *Cynegiticus*, or *On Hunting*, III, 3–7, translated by E. C. Marchant, The Loeb Classical Library, #183 (Cambridge: Harvard University Press, 1971), pp. 379–81.

will drive ahead, eagerly following false lines and getting wildly excited over anything that turns up, well knowing that they are playing the fool; others will do the same thing in ignorance. Those that stick to game paths and don't recognize the true line are poor tools. A hound that ignores the trail and races over the track of the hare on the run is ill-bred. Some, again, will pursue hotly at first, and then slack off from want of pluck; others will cut in ahead and then get astray; while others foolishly dash into roads and go astray, deaf to all recall. Many abandon the pursuit and go back through their hatred of game, and many through their love of man. Others try to mislead by baying on the track, representing false lines as true ones. Some, though free from this fault, leave their own work when they hear a shout from another quarter while they are running and make for it recklessly. When pursuing some are dubious, others are full of assumptions but their notions are wrong. Then there are the skirters, some of whom merely pretend to hunt, while others out of jealousy perpetually scamper about together beside the line.*

With all of this one may wonder how to go about trusting a hound, what to do in a real situation about one's dependence on the dog's honesty and character. One good answer is to train the dog properly ahead of time, but it turns out that in the situations where it counts, there really is no "ahead of time," there is just you and the dog. Bill Koehler tells a story about trailing a murderer in Fontana, California. Fontana is a pretty rough town, the home of the Hell's Angels, and we used sometimes to say that there were parts of Fontana where murder was a misdemeanor.

The dog involved was an English Bull Terrier. His name was also Bill, so to avoid confusion he was generally called Old Bill. Old Bill was a difficult dog to "read," and while in this case he was going along quite industriously, with every indication of sincere dedication to the job at hand, it wasn't clear

---

*A note on the etiology of character: Xenophon goes on here to say, "Now most of these faults are natural defects, but some by which hounds are spoilt are due to unintelligent training."

whether or not he was taking Bill on a wild-goose chase, which he wasn't above doing. He was frolicking unsuitably, for one thing, although that wasn't in his case terribly significant. The trouble was that he was taking Bill on a most unlikely route, through the backyards of a fairly sedate and well-populated neighborhood and away from paths Bill and the police would expect the culprit to take in a search for sanctuary.

Along the way, a Chow offered to engage Old Bill in combat. For Old Bill fighting was addictive and visionary—you have to have handled certain terriers to know what this means. The Chow was not only offensive in his language but of a size and strength to be a worthy opponent, yet Old Bill passed up the opportunity, an astonishing event, and, of course, I wouldn't be telling the story if they hadn't found the malefactor.

It was not Bill Koehler's ability to utter an imperative blindly and trust thereafter to providence or the dog's generous nature that enabled him to handle this wonderful and wonderfully difficult dog successfully. He had to be willing to chase a murderer for one thing, but mostly he had to be willing, in Stanley Cavell's words, to let his knowledge come to an end. Prior to that, there had to be talk in the first place between Bill and Old Bill, for there to be markable boundaries to the end of knowledge—dog training is not ethology. Tracking training creates the kind of knowledge all talking does, or ought to do—knowledge of the loop of intention and openness that talk is, knowledge of and in language. The dog as well as the human must learn this participation, which is why, if you put a harness and lead on your own dog and say, with all the humility in the world, "Go get 'em, boy!" you will find that ignorance is not enough.

Most tracking-dog handlers use the phrase I used when I said that Old Bill was a difficult dog to "read," in order to focus awareness on the handler's active participation in the conversation of tracking. It may seem misguided or misleading to couple "reading" and "conversation," perhaps because of the way books and newspapers, taken in isolation, can seem to be passively isolated, somewhat in the way tracking harnesses do.

But a picture of a person alone perusing a book is not a picture of a person alone; the book, like the tracking harness and the visual signals the dog sends, is a rhetorical connective, a metonymy. Reading a dog is no more a matter of one-way knowledge than listening is, as in, "She's a good listener." In this context the word "reading" is a word for a particular kind of conversation, a working conversation, that produces or invokes the knowledge of the conversation.

This reading, or conversing, begins with the human in command. There are various ways for the human to be in charge, and, as with any other area of animal work or human work, various ways the human can obscure from herself and onlookers that she is in fact in charge. The training method I will describe part of here is philosophically at odds with a number of others, in that a fair amount of time is spent setting up tracking as a purely obedience problem. This means that matters are arranged so that the dog is given opportunities to fail, and is corrected emphatically for doing so. I ought to say a word or two about that controversy—I am not neutral about it.

Roughly, the opposing way of thinking about tracking argues that, in the first place, there is no way for a human to correct a dog about failures to scent because we can't scent, and besides the dog has to *want* to track (or, if the author is contaminated by skepticism in some way, then the dog has to "want" to track). Tracking should be fun, so in the beginning you make a game of it. Both of the Koehlers, true to form, regard this way of talking as an insult to the dog's intelligence, saying that that is why it doesn't work so well. I agree with them—it is part of a general notion that being a dog is a matter of perpetual puppyhood, a demeaning notion, but also a natural enough confusion, a consequence of the fact that we really are, as I keep saying, in charge, and perhaps a consequence as well of the lack of any resemblance between a mature dog's way of being mature and what John Hollander has called "that gross travesty of maturity that passes for being grown up in America," the result of "putting away the wrong childish things." It is fortunately extremely difficult, though not impossible, to induce a dog to put away the wrong puppyish things, so a tracking dog in harness working with high seriousness also exhibits High Gaiety, the gaiety that goes with all true thought. This is perhaps why Koehler's central point about what actually motivates a tracking dog is so hard to see. He says that there are motivations more powerful than instinct, including the instinct to play games. Dogs, he says, *like people*, get the greatest satisfactions from doing something that is difficult well. But he is not so foolish as to suggest that difficulty in the abstract, difficulty not significantly anchored in the world, is a motivator, for most dogs at least. Using a knife and fork is hard for a dog, but there are very few of them whom that difficulty spurs to action. (Though I probably ought to confess that there are dogs for whom what is ridiculous is appealing, who would try to use a knife and fork just because it struck them as funny, as it strikes us as funny. Such dogs are good at comedy routines.) A dog who is track-sure is, most of all, undistractible. Pheasants may explode under her nose, or her worst enemy may offer to fight, she may become footsore, hot, cold or lonely, but if she has a true handler she will keep tracking. She is, by and large, so honest that most novice spectators can fairly easily learn to read her loss of track indications, even though very good dogs like Bill Koehler's Old Bill may be hard to read.

I am a creature who, for whatever reasons, is powerfully drawn to the idea of such a dog, so I decide to train Belle, my Pit Bull, to track, even though I have no way in the world of knowing how ultimately good her nose (heart) is; there aren't any prior criteria that will tell me, and the whole process of training may not even be proof, if she fails later, that it is because her nose (heart) is no good. The possibility that I have goofed always remains, though remotely in some cases.

Still there are signs. Her nose is well constructed, for one thing, even though she is not from a tracking line; more important, she is a responsible, single-minded dog. Also, she is protective. It may not be obvious what the willingness to attack in my defense has to do with finding lost children, and this is not the place to look in detail at this matter. But consideration of one example may be to the point.

I have not trained Belle in attack work. This is not because I disapprove of it. On the contrary, it is because I admire good attack work so deeply that I wouldn't do such work with an animal when there were any competing considerations, such as my being employed to teach English in a university. When you are handling a trained protection dog, that's what you are doing. Period. You can't also be monkeying about teaching the future leaders of our nation, or doing anything else peripheral.

Belle, nonetheless, is turning out to be quite a comforting presence, especially on the streets of New York. Not because she menaces people at random, but because she is continuously thoughtful about what is going on, and she stays on the job whether or not I do. In my office at school, she is quite peaceful. Students in all sorts of frames of mind, many of them dubious, can come and go freely, including students who are terrified of dogs, or of Pit Bulls, or of English teachers. Indeed, in the presence of a frightened visitor, Belle puts effort into radiating a quiet, reassuring sweetness, and so far as I know all of the initially nervous of my visitors have quickly become calm, usually wanting to give Belle a pat or say a word to her as they leave.

But one day, during some labor troubles, an anti-union or else a pro-union bozo approached me angrily, waving a document of some sort. Instantly there was a noise as though the placid old university building had become an active volcano, and the creep retreated rapidly, muttering about my politics.

This means that the experience of being in a university all day, with hundreds of people of all sorts coming and going, did not dull Belle's sense of what is what. She has convictions about how people ought to behave, and the courage of her convictions as well. This is the courage of thought, the courage to keep on thinking about things in the face of a situation that would lull a great many dogs and people into an accepting slumber. It is hard in the same way it is hard for a teacher to keep thinking about teaching, to keep letting it matter, and doing that well enough so that in a crisis it is still possible to remember to teach, to remember what teaching is. My experience of such crises, in the late sixties and more recently, have suggested to me that the number of human beings who can respond appropriately to a new situation instead of reacting merely with automated and suddenly atavistic responses is as small as the number of dogs who can persist in being genuinely, thoughtfully protective, the way Belle is.

The view of things that leads the trainer to call a protective dog either courageous or intelligent is one in which courage is a sign, or even a criterion, of thoughtful commitment to a job. When this courage is manifested in a tracking dog, it is so valued, and so essential, that there is no need for an adjective for it; once you know what a tracking dog is, the claim "Joe is a tracking dog," occasionally, "a real tracking dog," says it all.

What this means is that I can't just walk up to Belle and call on that courage. It has to be developed as any form of understanding does—there are disciplines of courage—and I have to earn this invocation, this calling forth of the powers of significance.

So I have to train Belle. She is already a retriever, willing to do a great deal more than dance out prettily after a stick. She knows her job and does it. And she has mastered scent discrimination and is in fact on a team of relay racers. In canine relay races, each dog plows out over a series of obstacles, picks up from a platform at the end one of four dumbbells—the dog must choose the one scented by her handler—and returns with it, carrying it home over the jumps.

In one race, just as Belle started out for her run, another dog, a somewhat absentminded Old English Sheepdog, meandered onto the course and stood lumpily between two of the jumps. Belle, without faltering in her stride, aimed her shoulder at the Sheepdog's flank, knocking him out of the way, and went on as though the path had been clear, doing her job. This is responsibility of a high order, not the sort dependent on cues of approval, from me or anyone else.

This deeply motivated retrieving is what I use to begin our conversation about tracking, rather than some fun-and-games routine where I "hide" and the dog goes bouncing along to find me. The latter sort of activity is part of a way of hanging out with dogs that is fun and that provides both dog and owner with a lot of healthy outdoor exercise. There is no harm in it, but it isn't dog training.

I begin in an area that has enough ground cover so that a small object, such as a glove or a sock, is invisible from Belle's point of view until she's quite close to it, but not so much cover that I can't see the scent object. I have an assistant, who will lay tracks for me. My tracklayer scents up a glove by putting it inside his shirt for a few minutes. I then take it from him and put it in Belle's mouth, saying as I do so, "Fetch!" I take it from her and praise her lavishly for accepting it, even though she is perfectly ready to retrieve much more difficult objects at some distance. I repeat this a few times, and then have Belle do an arm's-length retrieve of the glove, three or four times.* Then I begin the assignation of responsibility.

I stand with Belle sitting at my side, the two of us facing into the wind. I put Belle on a down-stay beside me and toss the glove out about six feet, wait a few moments and then send her with the "fetch" command. Despite the slightly unusual situation—she has so far always been sent from a sitting position for one thing, and for another, this is the first time I have asked her either to break her Down in order to retrieve or to pick up an object she can't see once it has landed—she goes at my command and brings back the glove.

Now I have her perform this routine a few times, only with my tracklayer standing at the end of the six feet. Instead of throwing the glove myself, I have the tracklayer, who is holding it, drop it at his feet, and I use a different object—simply another glove will do. Belle goldbricks a few times, and when she hesitates I "get on her ear" and pinch it. She carries me, screeching to high heaven, all the way to the glove.

Now we actually work a short track. While I stand back with Belle, the tracklayer lays a thirty-foot track into the wind, and when he reaches the end of it he turns and drops the glove at his feet. I put the harness on Belle for the first time but leave

---

*This description, like all descriptions in this book of training situations, is much abbreviated—you won't be able to work a dog from it.

the training collar and leash attached as they were. We come to the start of the track, where my assistant has scuffed up a spot a few feet square. I put Belle on a Down Stay with her nose in the middle of the scuffed square, and wait for two full minutes.

Two minutes is a long time. Belle checks and rechecks all of the scents available to her restricted area, largely because there is nothing else to do. There are, of course, a variety of scents in the area of her nose besides the tracklayer's, including mine and that of a squirrel whose headquarters are nearby.

The previous retrieves, where the glove was not much more than an arm's length away and detectable to the dog by air scent without the aid of a track, have told her that she now has to use her nose in order to obey the "fetch" command to find and retrieve the object. She has, of course, used her nose in the past to discriminate objects that were in full view, she has naturally used her nose a bit together with her eyes, but now she must, I hope, actually track, although the glove is still quite close and the tracklayer himself standing there at the end of this tiny track as big as life and twice as obvious, which means that the track is possibly the least apparent of the clues I am providing. This doesn't matter—if I have used the wind correctly she will travel down that beam of scent* to get to the glove, and what I am still concerned with is increasing the level of *retrieving* responsibility. I am not making it hard for her to use her nose. On the contrary, I'm putting up the dog equivalent of a whole bunch of gargantuan neon signs with big, brilliantly colored arrows on them pointing to the glove and saying: *It's here!* She may be learning that the track scent will lead her to the object I've sent her for, but she probably already knows that. The most important thing she is learning is that I am now going to back up the "fetch" command in a situation where she must use her nose without visual cues.

The next step is to have her follow these short, straight tracks into the wind without the presence of the tracklayer as

---

*As usual, I am using a visual metaphor for what I have no literal language for.

a clue, and I age the tracks about half an hour before I send her. Within a few days, I am having the tracklayer drop two articles on the track, so that Belle learns that her job isn't necessarily over when she finds the first.

The transformation from straight retrieving to actual tracking happens with astonishing speed. The first time I put her on the thirty-foot track, the response is unmistakable. Now she is moving out even faster and with more purpose than she had in retrieving, even though she retrieves quite passionately. When this happens, I change the command from "Fetch!" to "Find it!" I have waited until now to change the command because until now there hasn't been anything in the relationship between us that would make "Find it!" meaningful to *either* of us. I don't need to change the command, and not everyone who works their dogs in the way I am indicating does; "Fetch!" or "Take it!" can continue to be meaningful, to change and develop, the way other important words do. You don't need a new word for "book" when Baby progresses to the kind with no pictures. I just like the way the new command signals the new kind of thinking and imagining.

Belle's learning of the new command, or the more complete and commanding forms of the old one, is not unlike the way Stanley Cavell describes a child's learning a new word, say the word "kitty." Here is a fragment of his description:

Take the day on which, after I said, "Kitty," and pointed to a kitty, she repeated the word and pointed to the kitty. What does "repeating the word" mean here? and what did she point to?* All I know is (and does she know more?) that she made the sound I made and pointed to what I pointed at. Or rather, I know less (or more) than that. For what is "her making the sound I made"?... take the day, some weeks later, when she smiled at a fur piece, stroked it, and said, "Kitty." [Perhaps] ... she doesn't really know what "kitty" means....

But although I didn't tell her, and she didn't learn, either what the word "kitty" means or what a kitty is, if she keeps

*Or, with Belle, what does "finding the scent" mean? And what did I point to? Etc.

leaping and I keep looking and smiling, she will learn both. I have wanted to say: Kittens—what we call "kittens"—do not exist in her world yet, she has not acquired the forms of life which contain them. They do not exist in something like the way God or love or responsibility or beauty do not exist in our world; we have not mastered, or we have forgotten, or we have distorted, or learned from fragmented models, the forms of life which could make utterances like "God exists" or "God is dead" or "I love you" or "I cannot do otherwise" or "Beauty is but the beginning of terror" bear all the weight they could carry, express all they could take from us. We do not know the meanings of the words. We look away and leap around.*

So do dogs when confronted with words that do not "express all they could take" from the dogs, quite literally as anyone who is familiar with retrieving and tracking trials knows. When they learn the meanings of the words, have imagined the forms of life that give utterances such as "Find it!" meaning, they have not learned something *from* us, exactly—not learned something that we knew ahead of time. And with Belle I am learning something that is not something she knew ahead of time— although if I'm asked, "Who did you learn that from?" I would probably have to say that I learned it from Belle, just as I would have to say that she learned retrieving "from" me.

What are we learning? Trainers like to say that you haven't any idea what it is to love a dog until you've trained one, and there is a lot to this. When I first got Belle, I certainly loved her—in fact, I fell head over heels in love. I spent a fair amount of time just sitting and watching her, saying, "Oh, Pup!" And I said her name to anyone who would listen, as often as possible, the way lovers do.

Then I trained her in novice work, and when she started off-lead heeling something quantum happened, and "Oh, Pup!" became a phrase that compelled me anew, revised me. (Even though Belle is not the first dog I have trained, I still didn't know *this*.) With retrieving, "love" became capable of other

*The Claim of Reason, pp. 171-3.

powers yet, and now, in tracking, even on these ridiculously simple little tracks that Belle handles so easily that I might be tempted to forget myself and try harder tracks too quickly, I come to regard her with a new degree of awe and wonder, even though I am not *surprised* certainly at her ability to follow tracks so simple I can almost detect them myself when the humidity and vegetation are right. We are at this stage moving with some trembling into an arena where I will be wholly dependent on the dog's integrity to get the job done. Or, rather, an arena in which I can no longer escape knowing this about everything that is commanded between us; now this aspect of the shape of talking and loving emerges more clearly; that emergence *is* my new knowledge. Now it is something else again when I say, "Oh, Pup!"

As it is for her, too, as anyone can see. There is not only the pride and power of her dive down the track, there is her behavior afterwards as well. She has felt proud, exhilarated, in response to training in the past, but there is something new for her now. She responded to retrieving as a lot of dogs do, by getting her dumbbell out spontaneously and using it to encourage me to come outside and work, and all training increases her interest and proprietary pride in me. Now, in the house, instead of going to where the scent articles and tracking poles are stored and bringing me one, she waits until she has my eye and goes over and indicates them with a happy grin. If I say, "That's right all right, you're a tracking dog for sure!" or some such, she slaps a paw down on the pile of wind poles knowingly.

Despite this transformation, I am a long way from relying on her. For one thing, I still make the decision about when she starts the track and what counts as "the beginning of the track." Now, in fact (and with most dogs usually by the second or third time the harness is put on), she responds to the harness by beginning, without any signal or command, to look for a track, and when we approach the scuff marks the tracklayer has left, her head dips down to them and she moves to start the track immediately. I don't allow this but continue to make her lie down for two minutes at the start of each track; she doesn't go until I send her.

It is probably hard to see why, and even harder from this written description than it is in practice. At least, a friend of mine, watching us, wonders at some length why I don't let her track as soon as she wants to. As a matter of fact, with this particular dog I might almost be ready to begin responding *once in a while* with "Find it!" after she indicates that she has found the track. So this new "Find it!" (whether or not I begin that way of working at this point) is not exactly like a command anymore, or it isn't the same sort of command "Find it!" has been in the past. What I am after ultimately is a situation in which we have small conversations, agreements, about the moment to begin and other things. Later, and not so very much later, she will, for example, strike a trail, giving me some indication which can be hard or easy to read, and I will indicate, also with varying degrees of clarity, whether or not I want to go just now. This is complicated—something like the relationship between two dancers in a pas de deux when the decision is being made about the moment one dancer leaps for the other one to catch, or maybe something like the agreement between a conductor and an orchestra about when to begin—to say that the conductor is in charge is not to deny that the conductor must be responsive to the orchestra or that the lifting of the baton is only the most easily described of a web of gestures, utterances and so on that make the beginning possible. But in the end I am going to follow Belle more blindly than I suspect a conductor commonly follows an orchestra (though here I may be being stupid about music). I don't do this just yet. She still must hold her two-minute down at the beginning of the track, and I quite often choose a moment when she is obviously distracted from the work she is about to do by a wind shift that alerts her to passing creatures in the woods, by the noise of a passerby or by my cat Gumbie, who has a nearly infallible, though not, I think, malicious* sense of just when a dog is

---

*Cats are wonderful distracters of dogs in training. Some, like Gumbie, seem to be working out of a sense of cooperation. Others, for whom dogs have been offensive and invasive beasts in the past, are motivated at times by a somewhat more triumphant emotion, one can't help but suspect.

most vulnerable to the sudden emergence of a pussycat, and considers it her duty to test all dogs thoroughly. Gumbie strolls out, upwind. Belle's head and attention shift suddenly away from me, tracking and all that ought to be holy, and at *that* moment I say, cruelly, "Find it!" Belle, distracted by Gumbie, fails to respond and is ignominiously corrected while Gumbie sits and watches in a curious but disinterested way. "Find it!" is still quite fully a command, not a question.*

In the event Gumbie fails in her duties, I deliberately set up other temptations. If we have shrimp for supper, I save the shells, put them hot and tempting in paper bags and lay them two to three inches from the footsteps of the tracklayer. I behave in similarly unfeeling ways with bacon and hamburger, as well as with bottles of mountain lion, raccoon, deer, rabbit and woodchuck scent. If Belle were male, I would purchase essence of bitch-in-heat.† I am not satisfied until I am willing to bet a month's pay that if Belle hadn't eaten for two days and the track she was on led her directly through the middle of a platter of hot prime ribs, she would regard the food as a reminder to keep working, regarding it with impatience and disgust if anything. And when I reach the point of having a second tracklayer lay a fresh hot trail over the relatively cold one I want Belle to follow, I have him rub bacon fat on his shoes before laying the cross tracks.

While I have been exercising all this control over Belle, insisting that she scent what I tell her to scent and nothing else, and that she do it when I tell her to, I haven't improved at all, have made no progress toward being able to scent anything myself, though I can smell the shrimp and hamburger I set out on the track. So how can I know what I am doing or saying? How can I, even now, point at nothing as though I knew it were something and say, "Find it!" How can the "find" of the command refer to anything when I can't know the activity the verb expresses? How can the "it" possibly have a referent when

---

*It is not a question even with dogs who, like Belle, already scent-discriminate, thus answering the question "Which one is it!?"

†Literally, from, for example, the National Scent Company.

I stand a far better chance of learning to make my way from Canada to Mexico as migratory birds do than I do of ever detecting "directly" a short, easy, fresh track I have seen to the laying of myself? How, that is, can I ever know whether I am right or wrong to correct the dog, praise her, feel satisfied or not with the efforts of the day, and so on? How can there be any relationship at all between us?

(This is the sort of moment in my thinking when it seems to me clearly that syntax is prior to semantics—that you can't have meaningful communication without grammar—without a structure that is embedded in time. Even in the case of what appears to be a gesture or signal consisting of a single counter, as when Belle barks to be let out to run, the bark or tail-wagging or door-scratching is meaningless without what goes before and after coming in certain ways, in a rule-governed sequence. If she scratches at the door and then settles peacefully to her food dish, which is near the door, the scratching is neither syntactical or semantic. And relationships require syntax, as anyone who has tried to speak to someone having a severe schizophrenic episode is aware.)

Although I am not learning to scent as I work Belle, I am developing wind-awareness, and this is a curious thing. I learn, for example, that if a track is laid by a line of trees, the wind that I detect, by feel or by watching the flags on my wind poles or by seeing on a cold day which way my breath goes, may be in exactly the opposite direction from the wind at ground level. So I may be "sure" that Belle is going off-track, following something other than what I sent her to follow, when she is in fact working the air scent quite properly. If I am following a lost child or a criminal, and Belle leaps downwind to retrieve an object, thus claiming that the person we are following went *this* way, I will not necessarily have time to check out the wind in order to determine the truth-value of her claim. Besides, if the situation is urgent, I would rather hope that Belle wouldn't allow me to stop her in order to misread her. (One time a dog of my acquaintance, dealing with a handler who kept trying to turn him the wrong way, simply circled his handler twice, to build up momentum, and then took off at high speed in the right direction, yanking the handler off his feet; this is not an unusual kind of story, although its correct interpretation depends on knowing some other eerie things about dogs and handlers.)

So when I say that I am developing wind-awareness, I don't mean that my wind-detection powers have improved. But while I'm handling Belle I no longer believe my eyes in the way I normally do. I don't mean that I've gone blind, any more than Belle is blind, but that the general shape of my metaphysics has changed.

I still can't detect wind the way the dog does, I don't scent anything on it, and even if I did, the part that matters, down where Belle's nose is, is not something that I can detect. But what I have been doing, in the house with no tracking line in my hands, is studying books on scent, reading papers on grid patterns for finding lost persons in the wilderness, studying photographs of the patterns formed by smoke released by a "tracklayer," graphs of the changes in the weight and moisture content of a single fingerprint over varying periods of time and in various temperatures and humidities. This doesn't exactly teach me anything about scent, but it does alert me to how much my eyes aren't going to tell me, even though it is not wind-awareness.

William Koehler writes:

> The key to wind-awareness is not to memorize a lot of gerrymandering line drawings that picture scent direction relative to the track and the course a dog must travel to find or follow the scent. Those lines won't be out in the field to help you appreciate a problem your dog might encounter. But there's something that will always be with you. It's your ability to read your dog's performance relative to the air movement that careful checking shows you exists at ground level.*

So this is it. As time goes on, I keep saying, "Find it!" in different terrains and about different its, thus developing a

Tracking (manuscript), p. 41.

minute sensitivity to the dog's indications. It is in learning to believe her that I learn a particular kind of doubt of my own eyes, a skepticism that is a true "chastity of the intellect," as Santayana has it, so that eventually, on the actual trail of a child, when my eyes tell me that no five-year-old could possibly have climbed that rock face and that the trail must therefore go down that gully, I take due note of this. But I believe the dog and follow her. This is how there comes to exist for me, in our conversations about tracks, a kind of knowledge not possible in any other way.

This has implications for philosophy of language, real implications for a real philosophy, and if the notion of skepticism as chastity is properly distinguished from what Auden, also reading Santayana, called "tight-arsed old-maidery," then the implications are far more important than the mostly bogus issue about whether or not it is anthropomorphic to attribute mental capacities to animals that are different from the ones I attribute to my word processor.* Here is an example, from Donald Davidson, of a kind of thinking that strikes the tracking-dog trainer as specious:

> The methodological advice to interpret in a way that optimizes agreement should not be conceived as resting on a charitable assumption about human intelligence that might turn out to be false. If we cannot find a way to interpret the utterances and other behavior of a creature as revealing a set of beliefs largely consistent with and true by our own standards, we have no reason to count that creature as rational, as having beliefs, or as saying anything.†

There are no two ways about this. Professor Davidson here says as plainly as he can that we must never suppose limits either to our knowledge or to the capacity of the language we

---

*It is a mystery to me that mechanomorphism is so often felt to be more philosophically chaste than anthropomorphism; why the dog-as-companion is a more intellectually promiscuous idea than the dog-as-heat-seeking-missile is.

†"Radical Interpretation," *Dialectica*, 27 (1973), p. 324.

already speak to touch the rationality of anyone else.* Such
refusals are of course expressive of a human tendency, or at
least a Western tendency†—the kind of example I like to consider has to do with the ways the sexes talk about each other
when they are in states of uneasiness about each other. A great
deal of male behavior, especially male display behavior, reveals
a set of beliefs largely inconsistent with and untrue by female
standards. So women will, if kindly disposed, say that boys
will be boys, or talk of wild oats, or forgive the male as one
forgives a skunk for its skunkness. In more nervous moods,
the same phenomena lead to contempt and terror; both responses
have to do with a sense that the behavior in question is wild,
or automatic, or instinctive—not, in any case, rational. I know
less about the details of the promptings that move men, but it
is not uncommon to find men responding to female styles of
dealing with clothes or with men or with philosophy by developing similar ways of accounting for what does not translate
into a set of beliefs largely consistent with what is true by male
standards. I use such examples in order to say that Davidson
is perfectly correct about circumstances in which we have no
reason to count another creature as rational. But, as I have said,
having no reason to believe something is not the same thing as
having a reason to doubt it.

Working with tracking dogs does not, of course, guarantee
a consistent philosophy, but it does force anyone who succeeds
with it at all to learn to recognize the rationality of dogs that
occurs on the far side of the limits of the handler's knowledge—
though a handler may fail to learn this about the other sex or
psychologists or philosophers.

However wildly handlers may disagree about what motivates
a dog, what is fair or unfair, what it is to "train a dog to
track," they are unanimous in their rejection of a philosophy

---

*When I delivered this material as a lecture at the New York Institute for
the Humanities in the spring of 1985, one member of the audience said that I
was being "unfair" to Professor Davidson. But look at what he says!

†I am thinking of Emanuel Levinas' account in *Totality and Infinity* (Pitts-
burgh: Duquesne University Press, 1982).

that supposes that all rationality ends where their olfactory powers do. John Hollander has suggested to me a valuable simile for clarifying this problem. He said, "Well, isn't it like looking down on a person working their way through a maze in a case where you can see the person and you can see their goal, but you can't see the maze. Yet you are in charge." That is the situation in training, when you can see the goal. In a real tracking situation, you don't know that much, all you know the location of is the dog, though you might have some distant, inferential guesses about where the goal might be that the dog doesn't have, reports by witnesses and so on.

But when the human handler elevates inference, knowledge *that* there is a track, over the dog's knowledge *of* the track (which I can't of course characterize), the result when they come to work a track that is unknown to them is that nothing happens, and this nothing is one of the most embarrassing and revelatory of events. When I am working Belle, I have to watch myself continuously for a tendency on my part to enclose her and the terrain in the restrictive apparatus of my (visual) assumptions.*

Once or twice a week, she and I meet with a group of dogs and handlers who lay tracks for one another and provide the discussion and speculations that help make one's own errors perceptible. One time we were working in some abandoned vineyards, in an area several miles square. Our small company

---

*I'd like to emphasize the importance of understanding these matters at this point by reporting that Professor I. Lehr Brisbin, at the Savannah River Ecology Laboratory, has told me about some research he is doing on the nature of scent. He is finding that the dog's ability to recognize scent from one part of the body (the palms of the hands, say) does not enable him to recognize an object scented from another part of the same person's body—the bend of the elbow, say. And he doubts the ability of police-dog trackers to recognize by body scent a person standing in front of them whose track (shoe scent? instep scent?) they have been following. I have talked a little with Professor Brisbin and with Dick Koehler about training studies that would clarify the technical questions raised here. The philosophical reader will have to take my word for it that the problems involved, while adjacent to the problems I am interested in in this chapter, do not overlap the considerations in this chapter. Tracking-dog trainers who are interested in the adjacent training problems are urged to get in touch with me.

included two police officers and their dogs, who were working on scouting problems. One of the dogs was Packer, a fine, solid-working German Shepherd. His handler was Officer Riddle.*

Another handler, playing the part of a hiding suspect, left the immediate area to "hide" downwind from the dog-man team. Officer Riddle, when told that the suspect was hidden, emerged with Packer, put on the harness and line and cued Packer to search into the wind. Packer stiffened and started toward a line of cypress trees that had been planted some decades earlier along the edges of the vineyards as a windbreak. Officer Riddle followed him. Then Packer slowed, checked some nearby grape vines, working now across the wind and then with the wind to his back.

Officer Riddle, figuring that there was nothing in that direction a human could hide in and that Packer couldn't possibly be working air scent from that direction because that wasn't where the wind was, stopped Packer authoritatively and redirected him.

Packer, checked and redirected to work into the wind, began romping about, plainly playing the clown. The handler was advised by the supervising trainer to call the dog to heel, without any reprimanding. Fifty yards to his left, the "suspect" emerged from a low drainage ditch that wasn't visible to Officer Riddle. We were all silent, contemplating the implications of this breach of trust for Packer's future work.

In an incident that begins similarly, Dorothy, who was working a nice-tempered Sheltie named Silker, got confused about where the training track was. (It is astonishingly easy to do this, even with easily visible stakes and plain landmarks.) Silker moved out on her command, but less purposefully than

---

*Police-dog trackers, army scout dogs and the like are motivated in lots of different ways, and a great deal of the time the training is ruined, either by phony conceptions of behaviorism imposed by resident psychologists or by various popular fantasies about the dog's being within his own nature, his tracking nature, his hunting nature or his attacking nature. None of this applied in a serious way to Packer and Officer Riddle, as it happens.

usual, and worked off to the right of the direction Dorothy wanted to give her, moving in a somewhat vague, uncommitted fashion. Dorothy, however, did not reprimand Silker, who suddenly moved forward and pounced on the sock the tracklayer had dropped, waving and shaking it like the very flag of vision itself.

Dorothy was embarrassed, seeing at once that the reason the little dog's work started uncertainly was that she had been cuing Silker unconsciously. I asked her why, if she had been so sure the track was not where Silker was, she had allowed Silker to keep working. She fluttered, "Oh, I'm such an old goofhead that I figured I'd better think about it for a while longer, and Silker acted honest." Silker's future is perhaps brighter than Packer's is; Officer Riddle's commitment to a certain kind of psychic invasiveness is stronger than Dorothy's. Of course, he's young, and Dorothy, for all of her claims to being an old goofhead, has about forty years of handling on him.

I may be creating the impression here that a single mistake means absolute failure (as if there were a set of binomial either/ors in the dog that were being activated). This is, on one understanding of the metaphysics here, right and (trivially) wrong. It is right in two ways. Officer Riddle's mistake with Packer was very nearly the end of that project, partly because it was the result of a serious fissure in the officer's conception of what commanding his dog could and could not accomplish, the result not of momentary inattentiveness but of a general psychic imperialism. And it was a highly charged moment for the dog, the first time he had been given such a problem. Analogies have occurred in my experience with humans in sending riding students toward difficult fences for the first time. A fence that is 4′6″ or higher is to some extent always a crisis, so what happens in the jumping of such a fence tends to color and shape the future world for horse as well as for rider. When the student jumps badly merely because s/he didn't follow directions, then all I have to do is sharpen concentration and resolve. However, if something I say or do causes the pair to come into the fence off balance, then there will be more difficult remedial psychic work. Few people have the patience to perform such work, or know how to do it. In any event, one thing that will happen is that I will say, "I'm sorry, I was wrong." But this means something different when so much is at stake—this is not just the sphere of the moral life, where my intention to be right limits my responsibility for error, but like the logic of art, in which, as Cavell says, "our intentions dilate our responsibilities absolutely."*

In training tracking dogs you cannot escape the knowledge that the moral logic of teaching is like the moral logic of art; the student who is learning something real must table for a while the ability to respond in an ordinarily rational way to the teacher's "I'm sorry, I didn't really mean that." When you say, "Find it!" you have to mean it well.

In any event, on the day I have been describing, the dozen or so handlers with their dogs have been impressed by the morning's demonstrations of what goes into meaning "Find it!" While we were working our dogs, a group of children were riding their ponies around the fields, and now one of them, a little boy, was crying. He had come out wearing a brand-new jacket, but it got warm and he took it off, tying its sleeves around his waist. It was gone, and it could be almost anywhere in the several-mile area. The supervising trainer told us to take our dogs out, on no command at all, and follow them wherever they sniffed and wandered.

The vineyards were intersected by a kind of unofficial town dump, so there were plenty of items of clothing and other discarded objects scattered about. It was not at all clear what the dogs could be understood to be looking for, and that plus the enormous area involved led most of us to agree to search out of politeness rather than anything else. Perhaps the very fact that we had no expectations, and were therefore not cuing our dogs, helped. In any event, the jacket was located within three minutes by a nice young Shepherd bitch. This was an indication, among other things, of the dog's almost uncanny

*Stanley Cavell, "A Matter of Meaning It," in *Must We Mean What We Say?* (New York; Oxford University Press, 1976).

ability to detect what does not belong, which is what makes police dogs possible, despite the bizarre theories their handlers sometimes have about what they're doing.

When I began working unknown tracks with Belle, she displayed some wavering, as Silker did—evidence that I had been meaning both too much and not enough with my command on earlier tracks. Now that my utterance could no longer carry my body's arrogant mapping in it, Belle wasn't commanded by it in the same way, and in her wavering the landscape wavered for me too. This is a wretchedness, a revery of chaos I can imagine doing much to avoid. But because Belle is so beautiful it seemed to me that in the clear red and black of her coat, in her being put together and able to move in the particular way she does, there must be a picturing of the significance of the landscape. I was saved from believing my eyes by the new palpabilities that had come to me through the varying tautnesses in the line. My body, moved forward by those articulations, remained in that way adjacent to the dog's thought. And I saw that it was not that my educated heart-joy in her beauty must prevail, but only that nothing else would. So Belle, quickly enough, began to move out solidly again, restoring and reimpregnating my words. Meaning accelerated, and I followed.

In tracking I do not lose myself and become Belle; that would be a horror indeed, of the sort Poe wrote about. As Joan Dayan put it, in a discussion of Poe's "Berenice":

> In the passing of the mind continually over itself in this revery, all parts of life, spiritual and physical, exist only in their exchange, in the translation of one into the other. As the seen world becomes the world of seeming, the so-called world of subjectivity, so the mind turns into the very object once external to it.*

This, thankfully, has not happened. My mind has not turned into Belle in an invasion whose logic, Cavell shows, would

*In "The Identity of Berenice, Poe's Idol of the Mind," *SiR* 24 (Spring 1985).

eventually require her death and mine as I turned into her corpse.

Such invasions can come to seem necessary because we can refuse to know that we speak into darkness. There is a glow at the origin of speech, but our words carry farther than we can see. There are various ways of finding out something about where they have penetrated, the best one being to ask another creature. So speaking is always a questioning that wants to be a calling, an invocation, an appeal to obedience which may fail, as when the voice of Orpheus, who, Ovid tells us, "never before had called out in vain," fails to call through the vengeance of the Furies. And we may fail; the creature we ask may indeed respond, hear, obey, but we may then fail to obey. In fact, we usually do, and once that has happened there are no criteria that will take us back to the origin of our own words so that we can find out what has happened. In tracking, it is only in the dog's answering illuminations that you know whether you have said anything at all, or what you have said, and if the dog doesn't answer, then that is that, for the moment at least, for language.

There are certain entrancements and entranced certainties that tracking leads to, and it is worth understanding these because all certainty is worth understanding. But to understand them only in the terms I have given so far is not to understand enough of the ways certainty comes and goes. To consider this, I have to turn to horses and think about their ways of carrying us to understanding.

If scent is ineluctably "a queer thing," as one canny old dog trainer had it, so is the horse's sense of touch. This is a harder thing to observe, or rather, it is harder to observe the vast differences between our powers of knowing the world through touch and a horse's, partly because we can touch a bit better than we can smell.

Comparing the neural apparatus of the horse's kinesthetic powers with ours leads, like the comparative diagram of dog and human noses I started this chapter with, to a humbling realization that when we talk with horses, we are talking into a kind of intelligence we can just barely have any understanding of, partly because we communicate with horses largely through touch, and in particular through the lips and mouth. A horse's mouth, as the horse grazes, reads the plant life with exquisite accuracy (which is why it is so rare for horses to eat poisonous plants; they mostly do so only when they are very hungry indeed, or when the plants have been so tightly baled in with their hay that separation is impossible). And their mouths are very clever, so clever with the matter of locks on gates that it is the rare horseman who has had no occasion to be grateful that we have ten fingers and they only one mouth. Hence horses have continually to forgive us for what must seem to them to be extraordinarily blunt and clumsy communication, most of the time. Fortunately horses are, as that great American rider William Steinkraus has said, on the whole more generous than humans are.

The spirit of the rider, if s/he wants to be a genuine rider and no mere keeper of horses, a passenger at best, must match the horse's generosity and expand to this language, which means taking the risk of losing, for the moment at least, a great deal of the knowledge human grammars enable us to specialize in, just as working tracking dogs means losing some of the competence of the eyes. Some people can learn to remain articulate in the ordinary human way while honoring the horse's intricate kinesthetic language, but some can't, and this is a large part of why riding is terrifying, as it is a large part of what is scary about initial immersion in any alien tongue.

With horses as with dogs, the handler must learn to believe, to "read" a language s/he hasn't sufficient neurological apparatus to test or judge, because the handler must become comprehensible to the horse, and to be understood is to be open to understanding, much more than it is to have shared mental phenomena. It is as odd as Wittgenstein suggested it is to suppose that intersubjectivity depends on shared mental phenomena—he likes to talk about color patches—since we are primarily talking to someone, which pretty much guarantees little area of phenomenological overlap, even when I say just what you would say. So we are still poking at the edges of the biggest mystery of all about language—that it exists.

Consider the plight of the fairly green rider mounted on a horse—and the plight of the horse. Say that the rider has a small collection of skills but isn't very experienced, and that the horse, while he is pretty safely "broke," is far from being the reliable old chap you are willing to trust the safety of frail Uncle Albert to.

Every muscle twitch of the rider will be like a loud symphony to the horse, but it will be a newfangled sort of symphony, one that calls into question the whole idea of symphonies, and the horse will not only not know what it means, s/he will be unable to know whether it has meaning or not. However, the horse's drive to make sense of things is as strong as ours—call it "reason," in the way Hume had in mind when he remarked that reason is just another instinct. So the horse will keep trying but (mostly) fail to make sense of the information coming through the reins and the saddle and will be at best suspicious of this situation. (If the horse is anxious in a certain way about contradictions, s/he will be made more or less hyperactive or psychotic by it, but I am not assuming such a case at the moment.)

The rider will be largely insensitive to the touch messages the horse is sending out, but because horses are so big, there will be some the rider will notice and fail to make sense of, or else make the wrong sense of, just as the horse does. If the rider is working with the help of a good instructor and is very brave (smart), then out of this unlikely situation will come the conversation we call the art of horsemanship.

The reason the rider must be brave hasn't a lot to do with the danger of getting dumped, run away with or scraped off under a tree, although such fears may present themselves as emblems of the harder-to-articulate fear which usually goes: "But how do I know what s/he's going to do?" (Since s/he won't talk to me.) The horse will have the same fear—"This rider isn't talking to me!"—because the one thing they both know for sure about the other is that each is a creature with an independent existence, an independent consciousness and thus the ability to think and take action in a way that may not be welcome (meaningful or creature-enhancing) to the other.

The asymmetry in their situations is that the horse cannot escape knowledge of a certain sort of the rider, albeit a knowledge that mostly makes no sense, and the rider cannot escape knowing that the horse knows the rider in ways the rider cannot fathom. As an eventual consequence of this, many riders come to have a stance toward horses that is similar to the stance Jonathan Gash's character Lovejoy takes toward women when he says, "I know women pretty well, but I can't fathom them," which is adequate for getting along in a general sort of way, though it's hardly visionary knowledge. The horse's corollary to this is, perhaps, "I still don't know people, but I can't help but fathom them." Horses who come to such a stance are the generally reliable trail horses with what the owners may regard as merely a few quirks. Horses who are temperamentally unable to tolerate that stance are the stuff of which great bronc and outlaw stories are made.

Say that the rider is brave. S/he jiggles the reins, makes some sort of heavy, inarticulate movements with the legs, and the horse generally starts ambling off—but not in the direction the rider thought s/he was indicating, and rather rarely in the manner the rider had in mind. Some riders learn, through the process that begins thus, quite a few skills in the saddle, but never learn to respect and read the horse. Some of these riders, happening on extraordinary instances of equine generosity, win at fairly advanced levels of competition, but that's another subject. I am, of course, oversimplifying to some extent, since "skills in the saddle" can't be wholly separated from genuine articulations and responses, but this is a bit like using phrases learned by rote in a foreign language—rote learning, however elaborate, is not speaking. Human beings have greater capacities for rote learning than horses do, a feature of the situation that, coupled with the generosity of horses, makes a lot of inadequate riding possible.

What humans tend to do, of course, is to substitute rote learning of saddle skills for openness to being understood, and not always through a will to remain obscure; with horses as with tracking dogs, ignorance is not enough. The apotheosis of skill can take you to the threshold of art, but cannot take you over it.

One reason the horse's ability to fathom you is beyond your ability to control the horse has to do with a phenomenon I remarked on earlier: our skepticism is largely entangled in the visible, and that is the area where we are, most of us, pretty accomplished at knowing when to doubt and when to believe. We know how to check it out if we aren't sure whether or not our eyes are tricking us. (A usual method of checking is to appeal to another sensory system—to reach out and touch, for example, or listen.) We learn this through a language whose preferred system is visual. On a horse, though, until you learn not only to read what your skin tells you but also to be, as it were, kinesthetically legible yourself, you are deprived of that very skepticism that is part of the matrix of thought by means of which we learn to be certain enough, most of the time, for consciousness to proceed with a fair amount of confidence, which includes confiding ourselves to what we know in order to know it.

Our rider is nervous because these ordinary procedures of skepticism aren't sufficient. Her skin is telling her nothing at all—she knows that—and there is rarely any point in checking the next sensory system, because the skin doesn't know not only what it is reading but also what it is telling. And it is even harder for humans to read horses visually without knowing them kinesthetically. As for hearing—well, despite the sound tracks of cowboy movies, horses make little use of their voices.

An analogous example was suggested to me by a mathematician, who told me that in the beginning stages of learning a mathematical subject, the instruction to construct a proof is almost impossible, not because the student doesn't know how to check a proof for soundness, but because s/he doesn't know when to leave off doing this, when to accept that form of skepticism that allows the unknowable to stay where it is. And my friend led me to understand that, just as there are no flawlessly sound horses, there are no flawlessly sound proofs.

Hence, skepticism applied at random won't get you very far in mathematics, just as skepticism applied at random to horses, in an attempt to get beyond skepticism, leads you rather violently back to the ground if you keep it up.

So I am speaking about the moment when there are no connections, either logical or causative, between where you are and where you think you want to be, when you are in the Gap or the Void, the nothingness from which something comes. This is terrifying to most people, since we are creatures who become anxious if there is nothing there—the most familiar emblem of this is the terrifying sheet of blank paper facing the writer. This nothingness is the absence of language, for it is language that gives us the world: there is "no entity for us prior to a way of thinking and to thinking in that way."* And for us thinking cannot give us the world without language, which is why so many refutations of proposals that animals have consciousness are refutations of animal language, or at least of communications that have a syntax. This means that the picture of animal unconsciousness is a picture of a creature outside of time, an immortal creature in our terms, since for us syntax is the metric of time.

But I am assuming now that our horse and rider have managed not to perform evasions of the void and are facing it. Some riders are more gifted than others with the temperament that doesn't leap anxiously to fill the void quickly (by, say, spurring the horse into a gallop or choking the horse into paralysis—sides of the same coin of terror). They can allow the nothingness to remain in place, to sit still psychically, listening, brave enough to abandon the skills of being in what has been the known world in order to inhabit the world knowable to horses. Here, they start "hearing" the horse's skin, and in doing so become comprehensible in their own skins to the horse. This goes by funny leaps and (literal) fits and starts, just as when human infants learn (human) language.

The result is obvious to almost anyone, or is easy to make

*Robert Tragesser, *Husserl and Realism in Logic and Mathematics* (New York: Cambridge University Press, 1984).

obvious. An instructive film I saw once showed Colonel Podhajsky of the Spanish Riding School and a highly educated horse
performing some very complicated and beautiful dressage
maneuvers. The camera from time to time zoomed in for a
slow-motion close-up of Podhajsky's legs and hands. There
was, even in slow-motion close-up, no detectable movement,
that is, no visible cues or aids. Sometimes such riding is
described by saying that the horse and rider become as one
creature with one will, which captures part of it, but there are
other ways of talking about it. Riders whose grammar is contaminated by behaviorism will speak of making constant
"adjustments" of the horse's movement, anticipating with minute accuracy the horse's deviations. But a more traditional and
better way is to speak of the wonderfully rich and subtle
conversation that goes on in this sort of riding. That character-
ization answers to my own experience, and the experiences of
other riders who say such things as that it is important to
remember that any decently developed jumper knows more
about jumping than any rider in the world. This doesn't mean
that the rider's analytical capacities don't come in very handy
indeed, especially on a trappy course, but the rider who tries
to advise the horse about such matters without participating in
the horse's understanding and knowledge doesn't get very far
(literally). This is true of all talking that is talking, of course—
it is participation in knowledge.

It may go like this: Say that I am on a very fine, honest,
tough-minded jumper, and we are approaching a large and
difficult fence. I say, "Valor, old pal, we want to do this in
four strides, so shorten up a little."

He replies, "Yeah, that makes sense, but at the pace and
cadence I have established our chances are better if I come in
with three strides and take the fence a little long."

I might agree and leave it at that, or I might say, "Well,
but I walked this course with a measuring stick and you didn't,
and if you take this one long, you will crash the next, which
is very close."

And then if he trusts me, he'll say, "Okay, here goes!"

This is a deceptive rendering, of course, for when the skin's grammar is in unimpeded motion nothing so slow as the kind of language one can write down is needed. The skin is swifter than the eye.

Intimate acquaintance with the horse's knowledge and leading the kind of life that entails the continual reimaginings of horsemanship mark the faces of some older riders with a look that I have also seen on the faces of a few poets and thinkers, the incandescent gaze of unmediated awareness that one might be tempted to call innocence, since it is not unlike the gaze on the face of a child absorbed in Tinkertoys or a beautiful bug, but it is an achieved or restored innocence, and it is also terrible, the way Pasternak's face was terrible in its continuing steadiness of gaze. It is not false to call it the look of the life of the imagination, although in the next chapter I will call it the human capacity for heroism. It is a capacity most education seems to defraud children of. I mean "education" in the sense T. S. Eliot had in mind when he spoke of the eternal battle between art and education and of Blake as a man in whose life art had won. He said that Blake "knew what interested him," and that "this made him terrifying." Our stories and legends about horsemen, whether they are historical stories of figures of genius or children's tales of effort, courage and sacrifice for the sake of horses, are allegories about what it is to know what interests you, which is one of the reasons the passion for a life with horses is so powerful in this culture. It may also be one of the reasons (but only one—this is another subject) girls are so likely to become absorbed by the living allegory of horsemanship at just the age when their developing sexuality inspires the rhetorical forces around them to work harder than ever at distracting them from what they are interested in.*

The Renaissance horseman and thinker Gervase Markham

---

*I don't mean that boys are not so distracted, but I sometimes suspect that the allegory of horsemanship is one of the few in this culture that are available to young girls as an allegory of the courage to remain interested.

I have heard a number of people, vaguely inspired by Freud, speculate about the passion for horses that develops in young girls. As a trainer, I came to see that the real question was not "Why do girls ride horses?" but rather "Why don't boys?" This is not to the point of the present discussion, of course.

talked about the need to learn the skin's modes of thought by saying that beyond all of the skills of mounting, of signaling with the reins and so on, "there is a secret pleasing and cherishing of the horse with the bridle, which the rider must accomplish with so unperceiving a motion that none but the beast may know it."

In order to understand this, one must get past the notion of the bridle as an instrument of the kind of subjection that, in my experience, exists only in the fantasy lives of people who have bizarre notions about the nature of power. Abandoning such notions is essential to conversing with horses, and I think that the fear of abandoning fantasies of power, which may also be a terror of genuine power, is part of the fear of horses.

But I must return to the problematics that attend the dailiness of this conversation which is of so fluid and unperceiving a motion, since these are the problematics in which we all live, consequent on the way the unreal pours in on us continuously. When I was first trying to make sense of the ideas in this chapter, and in particular of the notion that there was implicit in the philosophy of horsemanship a pointer toward a new trick for skepticism to learn, I talked with Stanley Cavell about it, and he responded with a remarkable letter, in which he wrote:

> Is what your horses are telling you something like this, that what is preventing their being known is not too much skepticism but too little? If they said this to me, I would start thinking about it on the following lines: There is something about horses (no doubt something about the soul that *that* body is the best picture of—the ordonnance of *that* thinness of leg with *that* modulation of thigh underpinning *those* eyes . . .) that challenges as it were sooner than other cases the (skeptic's) idea that the problem of the other is the problem of knowing the other. . . .
>
> It is something about horses (what makes them nervous, what makes them crazy, what makes them shy) that sooner makes us wonder what we conceive knowledge to be.

This is right, there is something about horses that has led to our calling them the messengers of God, and there is something about knowledge that has led to the mythologies of the fierceness of horses, a fierceness that doesn't exist in the natural animal, but rather in what we seem to want to deny about knowledge. Cavell went on in that letter to suggest:

> There is something specific about our unwillingness to let our knowledge come to an end with respect to horses, with respect to what they know of us. (This is how I am taking your saying that we aren't skeptical enough about horses.) The unwillingness... is to make room for their capacity to feel our presence incomparably beyond our ability to feel theirs. ...
>
> The horse, as it stands, is a rebuke to our unreadiness to be understood, our will to remain obscure.... And the more beautiful the horse's stance, the more painful the rebuke. Theirs is our best picture of a readiness to understand. Our stand, our stance, is of denial.... We feel our refusals are unrevealed because we keep, we think, our fences invisible. But the horse takes cognizance of them, who does not care about invisibility.

The horse does not care about invisibility, at least not in the way that we do from somewhere around the age of ten, which is when, certain psychologists tell me, we create a persona, a social self, a mask through which the Other and I must mediate ourselves in order to talk, most of the time. The oddly triumphant appeals to the Clever Hans fallacy in some denials of horse consciousness have so far been the usual response of science and philosophy to horses' capacity "to feel [fathom?] our presence incomparably beyond our ability to feel theirs." The capacity to know our presence in this way has been, oddly, also the grounds for a denial of equine intelligence.* This seems to me to be a denial of our own knowability, a case of our making mysteries of ourselves with the aid of science and philosophy. The capacity for unmediated knowledge of us is not unique to horses, but because we ride them, because they *carry* us, it is particularly hard to avoid noticing not only that horses know us but that they know us without yielding their own volition, which continues to belong to the horse.

I find myself wanting to say that the dog's capacity to track and the horse's capacity to carry us—to know and carry exactly

---

*As in the uses made of the story of Clever Hans I discuss in Chapter 1.

us—tell me that true knowledge is that which springs (like a horse from the ground, the creature that can fly without wings) from the emptiness on the other side of knowledge, the true skepticism that really has gone far enough to be a genuine discovery of the mind's method, the method of doubt. (Of course, doubt's contribution to the mind is the continual imagining of the destruction of doubt. This is a clue about happiness.)

The skeptic's voice, which I have been having a harder and harder time ignoring, may suggest that in all of this I have been proposing an attitude of unearned belief in the rationality of anything that mimics rationality, or that I want to dismantle the project of discovering the differences between humans and other mammals. This isn't it, of course, and it is hardly possible for me to say strenuously enough how much I think we can learn about—oh—about the concept of time, for example, by examining meticulously the limitations of various animal concepts of time.

All I want to say is that the machinery of doubt needs to be reimagined; it is or has become as fruitless to doubt that the higher animals have minds as it is for me to claim to believe that the skeptic exists while doubting that the skeptic has a mind (though in black moods of impatience I do). I want the queerness of this to be felt. It is as though I were to say, "I believe that you exist, but I doubt that you have a height, because 'is tall' is a vague predicate."* That doesn't somehow help.

What *would* help would be for me to doubt how far my interpretation of your words can penetrate, and to understand thereby that my knowledge of you, my interpretation of your words, is not brain surgery. It is the articulation of the ways your words interlock mine, and it comes to an end somewhere outside of your skin.

---

*"Vague predicates" are the ones like "is bald" or "is a large number" that defeat our attempts to define them by locating exactly where they begin. Say that 1 is a small number, and so is 1 + 1, and so for any small number when you add only 1 to it, yet in this way you can get to 1,000,000,000,000,000, and so on. Or: Fred is not bald. If he loses one hair, he is still not bald, but if that keeps happening he will be.

5

Crazy Horses

There was once, not too long ago, a mare named Halla. Riders with international ambitions were interested in her, because she had an enormous talent for jumping. But there were, in the way the story has come to me, those who turned away from the notion of riding her in competition. For a horse to endure the stress of international competition, s/he must be deeply reliable, steady-hearted, tough-minded, and Halla was nervous, quirky, both very sensitive and very intense. She was, that is, crazy from the point of view of certain horse trainers.

Hans Winkler, a German trainer who didn't think she was all that crazy, began riding her. Rode her, in fact, at the Grand Prix in Stockholm in 1956 and in other equally important events. Riders watched, impressed by Winkler's ability to bring the mare's passions into that order we call "the art of riding"—an art that some thought had been closed for this mare by her "insanity." They spoke—in a way that will become not exactly false, but too incomplete to be helpful—of Winkler's enormous tact in "reassuring" the excitable mare.

I put the term "reassuring" in quotes not because I think it is wrong—certainly Winkler is one of the most tactful and reassuring riders who ever lived, and that certainly goes a long way toward accounting for his successes, not only with Halla but with horse after horse—but because I believe it was something else about Winkler that was important with this particular horse. A little information about certain philosophical battles that have from the beginning been part of horsemanship is in order. They are quarrels roughly like the ones I have talked about earlier, about how dog training should go. One side says that too much insistence on obedience ruins horses for dressage and stadium jumping because the horse becomes discouraged. Hysterical versions of this theory tend to bring in mention of Nazi Germany, the Emperor Caligula and other ignoble moments in the history of our race. The other side tends to talk the way Colonel Podhajsky of the Spanish Riding School does, a way that makes "understanding" and "obedience" virtually synonymous, as in remarks like "But first and above all the horse's obedience must be developed, or he will never learn to perform the advanced movements with intelligence and joy." Since most traditions within German riding are founded on principles of the latter sort, World War II was one of the events that inspired some English and American riders to reactivate, in their manuals, accusations against the Germans that were variations of the ones with words like "Hitler" in them, and such accusations generally included remarks about how especially brutalizing and upsetting obedience is to sensitive, talented horses. (In the late Renaissance, it was the English that were so accused, in slightly different terms, by the French. Quarrels about training technique are almost never about whatever the surface issue appears to be. And there is always some truth in the accusations, since no theory is a guarantee of good horsemanship.)

In the first round of the Grand Prix at Stockholm, Halla and two other horses turned in clear rounds. (This means that they didn't knock over any fences and didn't exceed the time allowed.) So Winkler and Halla were to be in the jump-off, which would determine which pair won the gold medal.

However, trouble had developed on course. The next-to-last fence was a big vertical, and Halla was coming in too fast and long,* but her tremendous talent enabled her to clear the

---

*Winkler told me, incidentally, that that was the only mistake she made that day. This is one of many details that made me realize he counted Halla's achieve-

fence anyway. And Winkler's talent enabled him to stay on her, clamping his knees hard while she jumped, with the result that in the air he pulled a groin muscle badly and was in sudden and severe pain. And there was the last fence still to come, a huge wall.

The usual responses to sudden pain are to collapse, vomit, take to one's bed, etc. What Winkler did was tell Halla to go ahead and jump the wall. They went clear.

When it came time for the jump-off, he was in such pain that he had to be lifted into the saddle. He explains, "They tried to do something about the pain, but—well, medicine wasn't what it is today." This was no case of Prince Hal vaulting "with such ease into his seat / As if an angel had dropped down from the clouds / To wind and turn a fiery Pegasus / And witch the world with noble horsemanship." Winkler was proposing to ride a Grand Prix jump-off when the use of his legs and back—which control a jumper—was virtually gone. You may more or less dispense with your legs and back while loping around nowhere in particular, down the beach or across a plain, but not over a Grand Prix course.

One horseman who was there told me that he had thought it was madness to contemplate riding a mare like Halla in that condition, madness to suppose that one could get so much as a decently calm working trot out of that particular mare with one's legs and back gone. The idea of expecting *any* horse to understand, and to be able to summon the tremendous concentration required by a powerful and trappy jump course by him or herself, seemed ridiculous. The notion of turning a nervous, unreliable mare toward those stressful fences when the rider's control was suddenly gone—horsemen speak of a horse's being upset when a rider in one way or another "drops"* a horse—

---

ment in Stockholm as of a higher order than his own—an attitude that often marks the great riders.

*The most extreme way to "drop" a horse, by the way, is to fall off, which is a paradoxical little poem about what, in some worlds, a commitment to horsemanship is.

made many riders there think that Winkler's ambitions had blinded him to what he was attempting.

"But," according to Winkler, "I had to ride the jump-off, you see, because the German team was falling apart, and we only had three riders left."* (I heard in these remarks echoes of Martin Luther's "*Ich kann nicht anders.*")

It was obvious that Winkler's legs were, for all practical purposes, "gone." All he could do was "go along for the ride," telling Halla which order to take the fences in and not much else. (I am oversimplifying a bit here; the point is how the event appeared to sophisticated horsemen. Imagine a race-car driver suddenly unable to operate clutch, brake and accelerator except in a distant, awkward and weak fashion.)

But Halla was a horse, not a machine. "Of all creatures," writes Gervase Markham, "the noblest." And Halla, as horsemen say, took her rider home. She won the gold medal for him and the team medal for everyone else. "We were able to do it," said Winkler, "because we were perfectly matched, and Halla was so sensitive and smart that she wasn't the sort of horse on whom you could get away with lurching around the course, bullying and coaxing† her into an approach. So all I had to do was just sit there doing nothing—she worked everything out." That is to say, the very qualities in a horse who in the wrong hands gets labeled "crazy" and "outlaw" are the ones Winkler values, attributing the success of that astonishing

---

*In the competition for the team gold medal a country may send four of its riders through the course, and the ride with the most number of faults is then eliminated from the account, so that the final scores are from the three best rides. *But* you have to have at least three rides for competitions such as the Nations Cup or the team medal in the Olympics. So if Winkler didn't ride the jump-off, the other two rides would be pointless.

†This, incidentally, is fairly frequently the case with mares, whether or not they are as sensitive and talented as Halla. You just have to do it right, or they get upset, giving you the opportunity, if you want it, to find out instantly from the psychic and physical biofeedback exactly where you are unbalanced and incoherent. This is one of the things I like about mares—their insistence and their continual and subtle demands. However, it is what some riders dislike about mares.

ride to them. (Incidentally, take note of the fact that this man could describe riding Grand Prix jump-off under tremendous pressure and in great pain as "just sitting there doing nothing." Also note that one of the reasons the extreme difficulties of this sort of riding are often invisible is that the hardest goal is, as in dressage, to achieve such a high degree of subtlety and power that it looks like "just sitting there doing nothing.")

William Steinkraus, himself a gold-medal rider and currently chairman of the United States Equestrian Team, prints a photograph of Winkler and Halla in his splendid book *Riding and Jumping* and in the caption remarks:

> Hans' wonderful accuracy in riding approaches enables him to leave the ground with little to do except maintain equilibrium, and his position usually shows more emphasis on delicacy of contact than on security.... This combination [of Halla and Winkler] has an astonishing record in winning the most important competitions, reflecting a very delicately balanced relationship between two high-strung and intensely competitive temperaments.

Steinkraus has here written a little allegory of virtue, or *virtu*—that noble old conception of intelligence, courage and power collapsed into one trait. In the language good writers and riders like Steinkraus use to talk about such people and horses and such moments, those figures of speech that are usually called "anthropomorphic"—as in the phrase "two high-strung and intensely competitive temperaments"—are transformed into such allegories. In our daily, secular language one might say that the word "temperament," which refers to innate moral qualities, belongs figuratively to animals and literally only to people. But in Steinkraus' caption, the allegory obliterates the gap between the literal and the metaphorical and becomes what the philosopher Charles Taylor calls invocative. Winkler's genius lay at least in part in his ability to be commanded by the invocative—to be commanded by the various poems of horsemanship in such a fashion that he called into "real" being, in a sense that even the realist might acknowledge as reality, his and Halla's heroic ride.

One way of providing part of the explanation of Winkler's success with Halla, then, is to say that he had a better story to tell himself and her about the nature of horsemanship and horses than riders who failed with her did. (Of course, our usual way of saying this is to say that he had a better theory.) He had to have had, for one thing, a story about how what appears to be horse insanity may be—even must be, most of the time— evidence of how powerful equine genius is, and how powerfully it can object to incoherence. And he must have had a story like that about Halla that would move him to keep thinking and trying when things got rough. Riders who have a lot of stories about incurably crazy horses tend to find that a lot of the horses they run into are incurably crazy. The stories we tell matter; stories not only reclaim the beauty of crazy horses, they may lead to insanity in the first place. I want to understand what a crazy horse is, and I strongly suspect that the epic size of the story of Winkler and Halla indicates that to understand any form of what we too often call craziness entails understanding that it occurs in an epic landscape, an epic world.

This means that the question "What is a crazy horse?" is incomprehensible if it is understood as a question it is psychology's job to answer. It is not psychology that has enabled anyone to do anything about the actual maniacs souring away out there in the corral, not psychology that has told the stories that restore horses to themselves and to us. The visions that have led to genuine horsemanship have been visions that come from artistic rather than psychological thinking, and even from mathematical rather than engineering thinking, since the great horsemen tend to have a sense of how form matters that is like the thinking that led to Plato's proof of the five regular solids. The first question is: What in the world could account for anyone's success in training horses at all, whether or not they are crazy? No description of technique, of the rider-as-subject, horse-as-object sort, however minute its details, can account for the myriad moments of transformation that are daily fare in the good trainer's world. Such moments are among the things that move trainers to turn with pitying exasperation from semi-Marxist descriptions of all training and discipline as coercion for the purposes of the state, which in this case is often represented by the trainer's "ego." In an age in which Marxism is not unbridled but merely enfeebled, this leads some writers who know nothing to speak of about horses to use horse training, not as it used to be used, as a trope of courage and genuine discipline, but merely as an example of coercion. So one finds remarks like the philosopher Foucault's that dressage is an "uninterrupted, constant coercion [that] by making the body docile . . . controls its forces, both to keep them from politically dangerous expression and to make them economically useful."

I do believe that things like education by and large serve to defraud humans of their own interests and sometimes thereby of their souls, and that crazy horses are one consequence of the "education" of horses. But as T. S. Eliot was moved to speak of "the eternal struggle between art and education," I am moved to distinguish between education as coercion and dressage or any other genuine discipline. Horse trainers turn away from such passages as Foucault's, and one way of expressing what they reject might be to say that it is simply wrong about the nature of dressage. But it might be better expressed by saying that it is ugly. I think it helps to remember that the horseman shares with the mathematician an ability to be commanded by beauty, even in the face of paradox, and for both the need to make the right distinctions between the beautiful and the merely pretty or picturesque is a condition of survival as a genuine thinker. This is embedded in the literary tradition, in tales of great horses who were disregarded because they were not fashionably pretty, and in certain places in artistic tradition. In Stubbs's portraits of great racers, for example, details of physique that would put a horse out of the running in a conformation class are carefully rendered, and in some cases, like the portrait of "On the Turf,"* there are details of posture, expression on the jockey's face and so on that express Turf's difficult temperament. Colonel Podhajsky, who trained the Lipizzaner Stallions, insists that "if the horse becomes more beautiful in the course of his work, it is a sign that the training principles

*In The British Art Center, Yale University.

are correct."* All by itself, this remark won't make you able to train a horse, but the rider who rests content with imitations of beauty finds his or her horses going crazy, although sometimes not violently enough for this to be noticed.

Beauty is a sign, even a criterion, of truth. But what is the craziness of a horse a warning against? This I want to understand by examining some of the ways the trainer summons the beauty of the horse. I don't know a great deal more about the history of Hans Winkler and Halla, so I will turn to my own experience with horses, one of whom was in temperament very much, I think, like Halla, although I don't know if Halla ever reached the extreme condition of fear and distrust that Drummer Girl did.

The standard crazy-horse story, which may have a title like *Outlaw Roan* or *My Friend Flicka*, or which may be a true story like the story of Halla, has a tendency to go in a certain way that seems to provide us with metaphors for maintaining clarity in the face of our difficulty with craziness in general, whether of horses, of people, or with what the French philosopher Jacques Derrida called the "fissure in the allegory,"† the crazing in the enamel of our understanding of the world.

In such a story there is an outlaw horse. Not a *wild* horse, but an outlaw, one who is for one reason or another outside of the order appropriate to the kind of creature he is. The outlaw, like some schizophrenics, usually has a Ping-Pong mechanism composed of fear and rage in response to people. There are fine tales, like Will James's *Smoky the Cowhorse*, in which we see the world from the horse's point of view, as a place that is orderly and meaningful except in the insane corners inhabited by humans.‡ Then there are brutal memories, which

---

*In *The Complete Training of Horse and Rider* (New York: Doubleday, 1967).

†In a series of lectures on memory and mourning, given in honor and memory of Paul de Man at the University of California at Irvine in the spring of 1984.

‡In that story, Smoky's first contact with humans comes when he is rounded up for the first time and branded, and James's narrator is at some pains to insist

the horse insists on by way of explaining and maintaining his craziness.

The horse's responses generally earn him or her more brutal memories, and there may even come a time when s/he is hunted down for use as a saddle bronc or as dog food. At this point, often, a child enters, boy or girl, usually in early adolescence. The child interposes himself or herself between the horse and the horse's persecutors in quite brave ways. Sometimes the bravery is largely a matter of spiritual or emotional stamina and clarity, but often there is a physical element—the child stows away in the boxcar the criminals are using to take the horse to be a saddle bronc, or simply places the tender child body between the rampaging roan and the raging men with ropes and/or rifles. The degree of violence varies with the cultural setting, but the threat to the horse is always absolute. In the hands of the right writer, the mere threat of being separated from the child is sufficient to make the horse's and the child's doom complete.

In one way or another, the child comes to have the horse available for handling, riding, training. (Sometimes the child steals the horse.) The details of this training vary enormously, depending again on which horse subculture is involved. (A variant of the crazy-horse story is the ugly-horse story, like the one Marguerite Henry tells about Hambletonian, foundation sire of the Standardbred, or the story of Snowman, both favorite American true-to-life stories.) A central detail in the training of the crazy horse is the child's courage in continuing to touch the horse tenderly, with both hands and voice, in the face of the horse's hostility. The horse, it should be noted, usually becomes dangerous at the sight and stench of the works of man only; with other horses, if they are elements in the story, s/he generally lives up to the horse's reputation as the noblest, most cooperative and forgiving of creatures.

---

that it is not the physical pain of the brand that is central to what is "alienating" about the experience, but rather the incoherence of it.

The child's touch penetrates the horse's confusion, fear and rage eventually, and *voilà*—the most stalwart and clever cow pony on the range, or the gentlest child's hunter in the park, or the greatest racehorse of all, as in *The Black Stallion*, or the harness-horse champion. Now, while I am on my way to denying that gentle touching is sufficient for any real wild-eyed rogue, it is certainly necessary. Horses can ask to be caressed and touched in various ways, but it still turns out that saying, "Please scratch my back," isn't by itself enough to tell one's companion how to do so; so the horseman who is faced with a more dangerous situation than we usually are when someone wants their back scratched, needs some rather powerful tales to transmute the horse's frustration with language into an acknowledgment of language. Frustration with language is, in the nature of things, precisely where one starts with a crazy horse.

The trouble is that the crazy horses have not by and large read *My Friend Flicka*. (A few have, and there are stories about that, too.) If they know some version of *The Black Stallion*, they left during the shipwreck scene, or they know only the part of *National Velvet* where The Pit is careening dangerously around town. They know a story about crazy horses, but they know the wrong one, or else they didn't get a chance or refused to read to the ending where things come out okay; and the trainer has to figure out how to get them to the ending, whereas the horses just aren't interested in the story anymore. The problem is to get them interested, a problem Shakespeare solved in part with a keen sense of the right bawdy jokes in Act I, Scene 1, but the trainers quite often don't even have that much agreement at their disposal.

Beyond which, if the horse does know the right ending to a sufficiently rich story and has tried it out on a human who doesn't, the difficulty is that much worse. The greatest challenges come from horses who have been beguiled by some confused version of behaviorism, or virtually anything from the matrix of academic psychology, instead of by a story that springs from artistic tradition; sometimes such horses might as well be severely autistic for all the talking that is going to go on for a while.

Drummer Girl was a young Thoroughbred mare, not quite four years old when I first met her. (The word "Thoroughbred," by the way, does not mean "purebred," as when one speaks of a purebred Collie or Doberman, but is, like Percheron, Appaloosa and Tennessee Walker, the name of a particular breed, the breed that is used for the form of horse racing most people are familiar with, flat racing as opposed to harness racing or steeplechase, and quite often for show jumping.) Her conformation was delicate and perfect—a textbook illustration of correct slope of shoulder, angle of hock and pastern, ratio of head to neck, neck to back, back to legs and so on. She was very slightly over at the knee, but this is so commonly a trait of stunning performance horses that it might as well be called a beauty. Her balance was wonderful, her movement true and clean; there was no paddling, no wobbling, and her stride was long, accurate, straight and powerful.

She was also, when I met her, ignorant, nervous and confused by—one can only guess—the bizarre activities that pass for training on and about California tracks so much of the time. She had papers from the Jockey Club and a tattooed number inside her upper lip, confirming the track history her papers detailed.

If she was nervous, she was not yet suspicious and angry; she was young and tender, a filly still, possessed as most horses are of a forgiving nature and willing to listen to the horse stories I had to tell. I found her in the course of a search I was conducting for a client, a man who had considerable riding skills and wanted a young horse to develop as a show hunter. (It is important to note that there wasn't anything wrong with this man as a rider—he was far from being a moral monster. In fact, I like and admire him a lot. In horse stories, riders who fail with horses like Drummer Girl are moral brutes, but those stories are allegory, not psychological realism.) I tried the horse out and showed her to my client, who was impressed, of course. He bought her and had her trailered to her new home, in a neighboring state.

A year or so went by. Eventually there was a phone call. The mare was causing difficulty—would I take her in training?

When I asked why a local trainer wouldn't do, since Drummer Girl now lived a one-and-a-half-day drive from my stable in the Southern California desert, I was told that four local trainers had already kicked her out of their barns. Since one of the stories I was determined to stick to at the time was that I could train *anything*, I accepted, and the mare was brought back.

As she was being unloaded, she twisted herself coming out, aimed her hind end at me and kicked viciously. I dodged successfully that time and discovered shortly afterward that she had kicked her shoe off and kicked so hard that the shoe went through the (admittedly not of the sturdiest) wall of the tack room. It is very unusual for a horse to go so far out of the way to attack a human, especially a stranger. In fact, most horses go to surprising lengths, even when they aren't especially well trained, to avoid hurting humans. The young, silly, confused but tender filly was gone. This was a full-grown mare, enraged, paranoid psychotic, violently uninterested in *My Friend Flicka*. She had, as it were, been to Vietnam. She was saying, of prewar civilization, goodbye to all that. She had had it with humans and their stories.

She had especially had it with kindness, any story about kindness. One of the many things that inspired her terror and rage was a soothing pat on the shoulder. She welcomed such gestures the way a rape victim might welcome a strange man's compliments on her figure, with the schizoid Ping-Pong, terror-rage-terror-rage.

I may have created the impression that simple physical brutality and distress caused her denials, but this isn't so, or at least it's oversimplified. For creatures with language, there are very few if any cases in which you can point to simple physical brutality isolated from a semantics of brutality, because in training situations there is always a relationship, exchanges of some sort, a matrix that creates the fact as well as the interpretation of physical pain, which is inseparable from the pain. (It is the syntactical and semantic context of surgery that makes it a different phenomenon from massacre, even when the patient dies.) Drummer Girl was, as it happens, unusually and genuinely sensitive physically, unlike Pennies from Heaven, a horse I will tell about in a moment who had simply learned a story about how to be a crazy horse that included the bit about unusual physical sensitivity. (This, like the reports I make about the temperaments of various animals, is not something I can prove that I know, it's just the sort of thing you do come to know if you work with enough horses.)

In fact, I probably caused Drummer Girl more physical pain, to the extent that physical pain can be quantified, than anyone else had. The significant brutalities for this mare had been linguistic, moral in the full sense. And Drummer Girl was temperamentally very different from Salty, the Pointer. Salty was mostly unimpressed with schizoid transactions, partly because she was born comfortable with the knowledge that human beings are by and large dopes, but also because going as hard as possible was central to her. A racehorse with a temperament like Salty's can survive the incoherencies of the track much better than a horse like Drummer Girl, for whom balance, symmetry and coherence were at the center of the cosmos. Hence the depth of her enraged despair when the world failed to provide the forms that would make manifest the movements she inchoately yearned for. She desired balance the way any creature desires its own nature, and this meant more to her than what we usually call love. The human heroine of *National Velvet* desired something like this (although she was perhaps more like Salty than like Drummer Girl), a "pleasure earlier than love and nearer to heaven." So you couldn't shift about casually, however gently, in the saddle while on Drummer Girl's back, even in order to stroke her neck while at halt, to praise her for getting in right, for in that small tilt the sun and the moon changed places and hunted each other down in her heart. In this she was, of course, a horse. It is innately *in* horses under tack to rage, or in the case of most family pleasure mounts, to retreat in varying degrees into dullness of response when there is no authentic movement and intercourse. In this case, Drummer Girl's enormous capacity for precision and elegance was the measure of her capacity for maddened refusals of any lesser communication.

What in the world was I going to say to this mare? A

number of options simply weren't available. If, for example, I were to approach her believing some febrile nonsense of the sort I quoted earlier, about training-as-coercion, then my beliefs would be manifest in my body as I approached the mare, and she, reading this, would plain old kill me, that's all, if she could find a way, thus underlining a remark of William Steinkraus' about the horse being the ultimate authority as to the correctness of our theories and methods.* It is wrong to talk about coercion, not only because that doesn't happen to be what's going on in places like the Spanish Riding School in Vienna, where horses are trained to the highest pitch, but also because it cannot go on with horses performing at high levels; horses have souls, and there is an inexorable logic consequent on that that makes coercing them into high performance impossible. I don't mean that you can't coerce horses, only that if you do, you will end up, if you are lucky, with a dull, unenthusiastic mount, or if you are unlucky, with a Drummer Girl emerging murderously from the trailer. Similarly, notions that make coercion out to be the central force in the world can't explain why some people become artists, make love richly and joyously or have authentic conversations. Drummer Girl's rage marks with exquisite accuracy the point at which both behaviorism and Marxism, as well as perhaps deconstruction,† must come to an end.

That first evening, I put Drummer Girl into a large pasture, where she could wander loosely and at will, getting the kinks of the long trailer ride out of her system, even though I knew what would happen, for, of course, the next day the first problem was how to catch her. (One speaks of "catching" horses in a pasture, even when they are eager and gregarious sorts who need no true "catching." It just means putting a halter on the horse and leading him or her out to be groomed or tacked up, or whatever is on the schedule.)

I approached Drummer Girl with a halter and lead shank in

---

*In his introduction to *Give Your Horse a Chance* by Lt. Col. A. L. D'Endrody (London: J.A. Allen, 1959).

†When I use these terms I mostly don't have in mind the writers whose work "founded" the theories. Marx was not a Marxist, and so on.

my hand. She responded to this by dashing violently around the pasture, kicking and squealing in case I didn't understand what she was saying, which was: "You can't catch me!" But, even though I didn't have a helper, I knew a fair amount about herding horses, so I managed to get her into a corner and was on the point of slipping the halter on. Whereupon she whirled, stamped hard on my foot with one of her hind hooves for emphasis and bolted: "You can't catch me because even if you did I'd probably kill you, because I'm crazy and you don't know *what* I might do!"

I accepted this version of things, since it was obviously, for the moment at least, true. I limped back to the tack room and emerged with a long light whip, called a longe whip, which is used for emphasis but not to touch the horse when the horse is being exercised on a longe line. (It is not a bullwhip; the distinction matters.) When she bolted again at the sight of me, I made no attempt to herd her or get her any closer to me. Instead, I said, "Okay. You want to bolt, let's bolt." So I snapped the whip and chased her heartlessly around the pasture. When she wanted to stop, I popped the whip and kept her moving. (I did *not*, it should be said, wear her out physically, nor did I touch her with the whip; both things would have been stupid, brutal and dangerous.) After a while she stopped and faced me, her stance and expression a real study. She had a new message. The message was: "Jesus! You're the crazy one! If you don't know any more than that about horses, I'll have to keep you calm! You're weird!"*

I responded by laying the whip on the ground and approaching with the halter and lead shank held out in front of me, by way of saying, "The way to keep me calm is to let me halter and lead you." The first few times I did this, she started her crazy bolting again, careening about like a firecracker. Every time she did, I picked up the whip and "chased" her. Eventually she walked toward me and stuck her nose out quietly, handy

*Since I wrote this, by the way, I talked with a psychiatrist who told me, "I get along fine with paranoids, because I get mad at them. But I get mad at them about something they can control."

for haltering. This was not coercion. Cornering her, lassoing her, but most especially sweet-talking her—all of that is coercion. What I did (it requires the sort of commitment to the horse that is not entailed in sweet talk) was simply to set up a situation in which she had available to her certain clear decisions. Sweet talk would have muddied things up with emotional appeals.

I should say very quickly, if there are any trainers or would-be trainers reading this, that this method will not work with most horses who don't want to be caught—especially range horses and mustangs. This was a response to a particular and unusual problem of appealing to what I was betting was the common sense of a very intelligent horse who had a particular history and a particular metaphysics. She had a certain story going about being caught, her own story, and since I couldn't simply toss that story out, I had to revise it. But horses have nearly as many stories about what "You can't catch me" means as there are horses who refuse to be caught. No method or impersonal theory relieves the trainer of the burden of judgment.

Although Drummer Girl was technically at least "green broke" and therefore ridable, I didn't ride her for a little while. Instead I longed* her. By this I don't mean classical longeing, where the handler stands in the middle of a circle around which the horse works on a long line, but rather as I longe dogs, as a preliminary to teaching the command "Joe, Heel!"† This was because, although I now had her on a halter and lead shank, and there was a truce between us, it was a very uneasy sort of truce.

What I did, roughly, was what I did with Salty—run in the opposite direction from the one she was going. If she bolted to my right, I went left. If she bolted forward, I did an about-turn and ran. If she went up, I went down, to China if necessary.

And, as with Salty, I gave her temptations and distractions. I set up a bucket full of oats, and when she darted forward to

---

*Pronounced like the past tense of the verb "to lunge."

†See description on pages 51–3.

grab a mouthful (which she did quite reliably, having been bribed a lot with oats in the past during her "You can't catch me" sequences), I turned about and ran, my momentum as usual yanking her head around. And she followed for the same reason Salty did—in order to stay near her head while dealing with this clumsy ignoramus of a handler.

About the fourth time my speed spun her away from the can of oats, the situation changed. She tensed to dart forward and then stopped herself (the lead rope was completely slack), turning over in her head the cosmic implications of my behavior. She got it right, snorted, stuck her left eye up to my right eye and pawed the ground furiously. Stories about a sweet horsie who couldn't help herself and needed only affection and gentleness were what had made her crazy, but they were the only stories she had—her only survival tools—and she quite naturally didn't want to give them up. So I didn't soothe her, nor did I put any pressure on the lead shank to "steady" her. Instead, I banked everything on her capacities as a moral agent. I turned and ran again, and because she was now distracted, not by the oats but by her broodings on my outrageousness, she didn't move quickly enough to prevent her head from being yanked round again by my motion. This was the moment, however, when things changed, and she actually began paying attention to me, like a creature whose behavior might turn out to be comprehensible if not pleasant.

With Salty, who was merely cheerfully untutored and not crazy, the longeing, by bringing the two of us into the posture appropriate to dogs and people, was by itself powerful enough to spark that within her that made her a dog, and to inspire quite a bit of trust. Drummer Girl, in learning to concentrate and keep her balance in order to turn comfortably when I turned, was also getting in touch with her own nature and her own power (as opposed to mere force, to use a distinction the *I Ching* makes at one point). However, she had at her disposal some hidden premises that saner animals don't have, so she wasn't ready to trust me so soon, although she was finding the notion of trusting me extremely attractive.

So attractive that within only four days of her arrival, I

went out to the stable in the morning with my friend Eleanor, who wanted to know more about animals in order to think more fully about her own work with disturbed adolescents. She and I lounged about chatting, so I didn't immediately take out the tack and catch Drummer Girl, who by now knew the order in which I worked the horses and knew that she got worked first. She was impatient for work and came over to the gate, pawing the ground in insistent appeal rather than in fury and bumping at the gate with her nose. So I left my chattering. When I walked over to her with the halter, she was eager, and when I was too slow for her she impatiently reached her nose out and stuck it into the correct loop. After our session was over, I turned her loose in the riding area, while my friend and I continued to talk about what was going on, seated on a table under a pepper tree.

After her roll, Drummer Girl came over to us and gently interposed her nose between us, nodding her head a bit peacefully to make sure we knew that she wanted in on the conversation. Eleanor, who had never ridden a horse and who certainly hadn't the skills for dealing with a homicidal Thoroughbred, and who was in any event rather frail physically, scratched the mare's neck and shoulder (a gesture which, remember, just a few days before had inspired impressive displays of anger). She chatted with the horse, with Drummer Girl practically in her lap, in perfect safety.

It is not hard to make this out to be a tale of coercion, I suppose; all you have to do is put the right adjectives in front of the word "halter," or suitably modify the word "eager" in the foregoing paragraphs. However, as I keep saying, it would be impossible to get the response I have described from the mare by using such a philosophy. Drummer Girl still didn't fully trust me, so even though she had come fairly quickly not merely to tolerate but to enjoy the work on the ground—which progressed to heeling and stays and recalls, as with dogs—getting on her back turned out to be quite another matter. Trust deconstructs itself and goes back to being truce by means of a curious etymology that ought to be in the Oxford English Dictionary but is not.

One day I mounted her. She flipped her head back immediately I was in position and broke open my eyebrow. Instantly with no pause, just as though she were a carefully raised and civilized Lippizaner at the Spanish Riding School, I asked her to perform a volte at trot, instead of gingerly moving off at a walk. A volte is a graceful circle that the horse can't maintain at all without a fair amount of concentration.

Drummer Girl was so startled by this outrageous, unorthodox response on my part (remember that she had been in the hands of a number of trainers and so knew something about orthodoxy) that she gave me two and a quarter nice circles before she broke out again in upset. But her resistance changed, quite suddenly, from what a Marxist might admire as revolution to a halfhearted kind of objection that was more a function of habit than of story or philosophy. This was because her own beauty and talent had given her, in response to my request for the volte,* a trot that was just about classically pure for just those few moments. For a few steps, then, she achieved under saddle what horsemen call "self-carriage," and those moments of congruence and contact with her own splendor accomplished more than years of people babbling about her sweetness and prettiness had. She was in any event neither sweet nor pretty. She was, however, beautiful. (The sort of beauty that is or, as I would like to emend tradition, can be the beginning of terror, since to know beauty is to know the loss of beauty and thus full angst in the face of the knowledge of death.)

That it was beauty and not prettiness mattered, and continued to matter as I worked with her; the discipline of horsemanship entails continually making the right distinctions, not allowing oneself to be seduced, whether by the horse or by one's own flabbiness of soul, into either asking for or accepting merely pretty movement. Offering merely pretty movement was probably how Drummer Girl kept other riders at a distance from her, but that distance she knew so well how to maintain

---

*It is important to remember that I knew she had already had experience of the volte, which is not a movement to demand of a horse who is both green and crazy.

was the exact distance from which nagging is possible. So in refusing to nag her I was insisting that her movement be *true*, while remembering that this was a truth that came from her and not me. My job, as Wallace Stevens said of the poet's job, was simply to recognize it when it came. Which is, of course, yet another of the hundreds of translations and retranslations that horsemen have been making for five centuries of Xenophon's remark that in riding the development of the horse's beauty is the sign of a true training philosophy. But I haven't yet made clear what a terrible and wonderful thing the understanding of that remark, as worked out in the saddle, is.

Various things happened in the course of our work to complicate and change what I had to do and know in order to elicit from Drummer Girl her own power. When I first tried her out, one of the things that drew both me and her owner to her was her jumping. She was one of those horses that take to jumping naturally; she was in fact less nervous about fences than she was about work on the flat. And her form over fences was a thing to see; without instruction she brought from herself knowledge of all sorts of things about approaching and negotiating fences that normally I wouldn't expect of a horse for some time. She just liked to jump.

But she had become a fence rusher, so generally recognized as one of the hardest problems to deal with that a fair number of excellent training books advise people to pass a fence rusher by and try another mount. Instead of going forward eagerly and thoughtfully into a jump of whatever height, including a symbolic pole laid on the ground, a fence rusher pushes the panic button and charges the fence in a high rage. This is dangerous for horse as well as for rider.

I've seen and read various solutions for fence rushing. One method that once in a while works tolerably well, at least in terms of bringing home ribbons from local shows or enabling a rider to survive a fox hunt or a cross-country run more or less intact, is to put a lot of head gear on the horse: martingales, which hold the head in and down, and bits, which through leverage not only punish the mouth, chin and poll when the head starts to rise but also reduce the amount of pressure the rider must exert to hold a horse in, and so on.

Another method is to work the horse in circles, one of whose arcs equals part of the arc to the approach, over and over again, turning the horse from the fence ten to forty times for every time the horse is allowed to go on and jump. The hope is behavioristic: the horse will come to associate the sight of the jump with the quietness of the circles. I have seen this method work on horses (and on riders) I didn't consider genuine rushers—they were only temporarily unnerved or inexperienced.

Another method that works on sane horses or fairly experienced ones is simply to stop the horse on the approach to the fence if s/he starts to charge, and then negotiate the approach again until the horse comes in quietly and thoughtfully.

But Drummer Girl was dead serious about rushing. She would tense up and tremble as though she were having delirium tremens if, when we walked into the riding ring, there was any evidence of jumping timber at all. I estimated that we could circle round and round for the next twenty years, and it would serve only to reinforce her dedication to rushing. Martingales and various kinds of curb bits are, in the right hands and on the right horse, useful and even noble devices, but at this stage Drummer Girl was still capable of such an intense terror of being captured or hemmed in that I knew I had to rely on my own quickness in giving halt corrections rather than equipment that would otherwise have considerably reduced the amount of sheer physical work I had to do.

After suitable preparation I needn't detail here, designed to ensure that she understood what was going on, I made no attempt to slow her down on the approaches to the fence. (The "fence" was, at this stage, literally nothing but a pole laid on the ground, so the athletic problems she had to solve were virtually nonexistent.)

She charged at a pace and with a fury that made the Kentucky Derby look like a pleasure ride, and I let her go, leaving the reins entirely alone so long as she bolted over the pole. Trying to slow her only fired her up anyway; no one is strong enough to hold in such a horse, at least with the head gear I was using.

I couldn't slow her, but I could stop her. So, one and a half or so mad strides after the fence, I sat up and very quietly asked with my legs and seat—not with my hands—for a halt. She, of course, paid me no attention; one purpose of fence rushing is to try, in a kind of distorted *furor poeticus*, to revise the rider and the whole situation out of existence. Drummer Girl was not only doing that, she seemed willing to let herself be so revised, anything to avoid the by now terrifying business of thinking about jumping. I responded by giving her a violent halt correction, dropping the slack into one rein and using it to compensate for my relative frailty, jerking her head up and back in a way that brought her, perforce, to a sudden halt.

Since she understood a great deal now about the halt correction on the flat, this got her attention. We went back and forth, back and forth, over the tiny rail. I continued to leave her free to rush as she pleased on the approach, insisting on nothing but the halt afterwards. In time—not a great deal of it—she started to slow down on the approach and think about it, not because she had returned to her former innocent enthusiasm for jumping but in order to be *balanced* in such a fashion that she could stop when I asked her to after the fence without either getting the correction or having to stop by letting all of her weight slam painfully onto her fragile and sensitive front legs.

Soon she started, with no advice from me, to teach herself to come in with that thrillingly round canter—full collection at canter—that eliminates gravity, achieved on her own in the course of solving for herself the problem of dealing with the fact that she must stop after the fence, a traditional movement that is sometimes criticized by those who don't understand it as "artificial." She rounded her back, flexed her head and neck, flexed hip, knee and hock and brought her hind end up under her for greatest control and suppleness.

I had heartlessly set things up so that, if she was capable of any sort of reasoning at all, she must slow down and canter properly. Must, that is, choose a muscular version of the good over the chaos of the stories about a sweet, dumb horse who just couldn't help herself. I could do this only because there was a vein of harmony in her to call on, and once she had made *this* decision, she suddenly had within her the power to make further such decisions in other situations, even though I was still far from being all the way home with this extraordinary mare. She knew now what the good felt like, even if she didn't yet know many of the forms it takes in the uneasiness of actuality.

There are a few things in all of this I haven't said much about. There was, for instance, the fact that my first ride on this mare was made difficult for me because my eyebrow was split open, and there was a fair amount of blood stinging my eyes and interfering with vision. I did not get off to check the damage; that would have been a failure on my part to go forward, and going forward is a sacred value the rider as well as the horse must submit to, or the horse probably never will. Beyond which, there were risks for the horse in the procedures I am describing. Not anything like the risks young horses encounter on racetracks, but risks, and this must be understood without appeal to some calculus of suffering.

The way to avoid pain and risk for the horse would, of course, have been to give her a nice stall with access to enough pleasant pasturage to inspire her to move about and stay relatively healthy, instead of training her. The question "Why train horses at all?" is like the question "Why should we like the world?" Most of the time, the only answer that can be given is a maneuver rather than an answer, as when Wittgenstein in *Philosophical Investigations* said, "We like the world because we do," a remark that can be interpreted darkly or not, as you like.

If, for example, there was someone who wanted to challenge my involvement with training in general, and especially with horses like Drummer Girl, on the ground that it is "suicidal," then I might say, with some impatience, "Yeah, but you don't know how self-destructive I would be if I weren't doing this!" That, of course, wouldn't answer anything, though it might get someone off my back. I have said this sort of thing, and in fact had occasion to say it during my first ride on Drummer Girl, as there was a nervous spectator on the ground, yammering about emergency rooms and so on, and I said that I didn't dare get off yet or the horse would eat us all. But even if such an answer were true, it wouldn't tell anyone anything about what is going on. (Even if Freud were right, in a given instance, about why girls ride horses, that doesn't tell us anything.) I had fallen in love with horse training, and with this horse, and while love is a dangerous guide, there are parts of the forest we sometimes find ourselves in that no other guide even guesses at the existence of.

Horses themselves are not capable of forming the question quite as we do, but they are capable of seizing on a solution when they think they have found one, as I was taught by a horse named Pennies from Heaven. I don't know for sure why so many crazy horses have names like Pennies from Heaven or Sunshine or Blossom, but I can guess, as I hope the reader also can by now. Pennies was a thirteen-year-old gelding, a nicely but somewhat coarsely built Quarter Horse with a deep copper coat. Unlike Drummer Girl, he gave no evidence on the ground of being difficult—he trailered quietly, walked easily on the lead, was cooperative about being tacked up and groomed. And he didn't look crazy but seemed rather to regard the world peacefully out of large, serene eyes. He also, when I first saw him, was wearing the only baby-blue roping saddle I have ever seen.

Once anyone mounted him, he was a textbook maniac, though even that, despite his hard-core bolting, was deceptive to anyone watching from the ground; you really had to be on his back in order to feel how deep and persistent his insanity was. And he had, of course, been in the hands of numerous trainers. One of them had tried to teach him to stop by letting him gallop past a hitching post and leaping off as she tossed the reins over the post. So Pennies learned to slow down near her hitching post, but otherwise this vaguely behavioristic fantasy of training served only to teach him that perhaps he could get rid of the rider by heading for whatever resembled a hitching post.

Pennies was in any case easy to handle on the ground. I would lead him out to the center of the riding area, safely far from fences for crashing into and trees to climb, get aboard and start thinking and riding very intensely indeed. I mounted at a point that was also the crossing point, and thus for some exercises the stopping point, in a large figure eight. I could make Pennies stand still while I mounted, but once I asked, however lightly, with leg pressure, for a walk, he bolted. Here, as with the jumping exercise with Drummer Girl, I made no attempt to slow him but only insisted that he bolt in a figure eight, and that his bolting be interrupted by a halt every time we went through the X in the center of the figure. He, like Drummer Girl, started to slow down, collect and think, in order to be ready for the next halt command.

The first day, it took nearly two hours (it was August, and this was inland Southern California, a desert area where temperatures of over 100° are common) to get him to halt in response to the leg aid or request without a correction. The second day, though, he was halting about half of the time when I asked him to, and I gradually began to ask him to halt at other places on the figure eight besides at the X and to add to the demands I made on him in other ways.

At the end of two weeks he was calm at walk, trot and canter, at least in the confines of the ring, but two weeks wasn't much against his thirteen years of studying how to be a crazy horse, so when his owners showed up I had no notion of putting anyone but myself on their sweetheart. When they first arrived I happened to be working Pennies, and as we trotted by their observation post on the ringside fence, they called out that that was sure a nice little ol' horse I was riding, and if they waited for a while, would they have a chance to see Pennies work? That is to say, they didn't recognize their own horse, whom they had owned for a decade, so changed was he. The son, who was about sixteen, fancied himself quite a wrangler and wanted very badly, once he knew that this was Pennies, to ride. I thought that it wouldn't hurt if he just went about a little at a walk, and said this, and Andy got on.

At first he was still afraid of the horse and followed my

directions exactly. I allowed him to progress to a trot on a long rein. Andy had never trotted the horse before (Pennies had had only two gaits before, under saddle at least—halt and bolt), so Andy decided that the necessary transformation had occurred and he could now proceed in the old way. When I felt that the pair was doing well enough to try a little cantering, and had explained the leg aid for the canter, Andy proceeded to stick his foot out preparatory to a bold kick with a sharp-heeled cowboy boot. In the mini-second during which I read Andy's intentions I despaired, being quite sure that the kick would land both of them in the next county and me with a lawsuit, if not a criminal charge, after they scraped the boy's body out of some chain link fence.

Pennies, though, had been thinking about things. When Andy kicked him, he merely broke into a slow, shuffling, plow-horse-babysitting-the-kids-style jog trot. I shouted Andy to a halt and told him to give the aid correctly this time. But again he kicked, and again Pennies did his shuffle-foot routine.

This time I assured Andy if he didn't do as I said, I would use the bullwhip on him, and he was apparently still afraid of me even if he no longer was of Pennies. He obeyed, and Pennies surprised me, even though I had been reading him correctly, by picking up a beautiful, classical canter, with no help from Andy, who couldn't ask for that canter because he had never known it.

All sorts of risks were being taken here. Initially, for example, I was risking the horse's health and my own life by riding at all. But notice that in making his sustained effort to say to his family, "I know another way for life to be, and from now on I'm going to insist on it!" Pennies was also taking an enormous risk—call it a psychic or metaphysical risk—for the brief version I had had time to teach him of the story of the grown-up, responsible, in-control-of-himself horse was really only a kind of small sketch for the sake of which he was willing to give up a gigantic mural in oils by which he had lived successfully most if not all of his life. It had earned him a lot of soothing pats on the shoulder if nothing else, plus a lot of free time because, of course, he hadn't been ridden much in recent years.

The continual presence of risk in any kind of training tells me that understanding what risk is, what kind of logic is involved when we speak of a "justified risk" and so on is important. What you risk reveals what you value, or as Heidegger would have it, the grounds of your being. By "the grounds of being" I don't mean anything terribly abstruse—in my case, for example, horses and dogs form an important part of the grounds of my being, which can be seen by anyone who takes a ride through the countryside with me. My eyes tend to fix on any horses, dogs, ponies and so on in the landscape— they are the means by which I know the landscape. Someone else would read and know the landscape differently, by the buildings in it, say, if s/he were a historian of architecture.

I'm not sure that I've made clear how hard the question about the nature of risk is, except perhaps to the horseman reading this who will know that when I "chased" Drummer Girl with the longe whip, I was taking an enormous risk with her sanity that could, for all I knew, have been what flipped her into unalterable psychosis, the loss of all possibility of trust. In this case I made the right decision, the risk paid off handsomely, but it needn't have. I could have been wrong, dead wrong, as we say.

To indicate how huge I think the issue is, I am temporarily going to leave the riding ring and go to Vietnam, which is for my generation the ultimate emblem of senseless risk. At the end of Michael Herr's book *Dispatches*, a documentary account of journalists in Vietnam, there is this passage:

> One day a letter came from a British publisher asking him to do a book whose working title would be "Through the War" and whose purpose would be to once and for all "take the glamour out of war." Page couldn't get over it.
>
> "Take the glamour out of war! I mean, how the bloody hell can you do that? Go and take the glamour out of a Huey, go take the glamour out of a Sheridan... Can you take the glamour out of a Cobra or getting stoned at China Beach?

It's like taking the glamour out of an M–79, taking the glamour out of Flynn." He was pointing to a picture he had taken of Flynn* laughing maniacally. ("We're winning," he'd said, triumphantly.) "Nothing the matter with that boy, is there? Would you let your daughter marry that man? Ohhhh, war is good for you, you can't take the glamour out of that. It's like trying to take the glamour out of sex, trying to take the glamour out of the Rolling Stones." He was really speechless, working his hands up and down to emphasize the sheer insanity of it.

"I mean you know that, it just can't be done!" We both shrugged and laughed, and Page looked very thoughtful for a moment. "The very idea!" he said. "Ohhh what a laugh! Take the bloody glamour out of bloody war!"†

The speaker in this passage, Page, is at the moment in the hospital in Saigon for something like the sixth time, as a consequence of a piece of shrapnel in his head that ought to have killed him. For him and for the narrator, the only alternative to the "glamour" he is talking about is pallid luxury, a kind of life-in-death. And Flynn, the man whose maniacal joy Page celebrates, is horribly dead by this point in the book, as are most of the author's buddies. If it is possible to earn the right to say that you can't take the glamour out of war, Page and Herr have earned it.

Horsemen share with Page this much at least: the conviction that to turn away from risk is to live at best a life of pallid luxury, though I've not met many who thought that risk in and of itself had any value. But the horseman's relationship to risk is an example of a frequent sort of human response to the knowledge that we die, and to the way it turns out that once death has gotten into the imagination, not even immortality will get it out again. (Think of Tennyson's poem "Tithonius" in which Tithonius, Aurora's lover, to whom the goddess has granted immortality but not eternal youth, pleads with her for

*Errol Flynn's son.

†Michael Herr, *Dispatches* (New York: Alfred A. Knopf, 1977), pp. 265–6.

the restoration of his mortality. His plaint is one of a long series in our literature.)

I don't know of a trainer who hasn't had experiences of the sort my partner had one day riding a horse who was suicidal and homicidal in a more organized though less committed way than Drummer Girl was. (I suspect that this horse wasn't smart enough to achieve the kind of commitment Drummer Girl had.) He asked me to be on the ground whenever he rode the horse because, as he said, there ought to be someone there to get the forklift out and scrape him up from time to time.

At one point he said savagely, after a particularly malicious effort on the horse's part, "You know what I'd do if this were my horse? I'd kill him, that's what." He recovered his trainerly philosophy fairly quickly after this lapse, but he really meant it, this man who loved horses as he loved his soul and most of the time gladly took informed risks for the sake of their beauty. And this, of course, is a success story, a story about the success of my partner's imaginative courage, but it is also a story that needn't have turned out this way. There was every evidence that this horse was a genuine rogue, and it did look for quite a bit as if destroying him was the only safe thing to do.

William Steinkraus, who knows quite a bit about risk, has this to say:

> The only horse to whom the above recommendations do not apply is the rogue—the equine equivalent of the congenitally criminal personality. As in the case of humans, some rogues are so gifted that we are tempted to take the enormous pains they require in the hopes that they can be redeemed. Almost by definition this is impossible with the real rogue, and as a rider's experience expands he will begin to recognize the telltale gestures and expressions that distinguish the rogue, or the genuinely dishonest horse, from the variously spoiled and frightened horses that present some of the same symptoms. When a true rogue is encountered, the best advice is probably to dispose of him, no matter how beautiful he is, or how wonderfully he can go "when he wants to." However, *such*

horses are very rare—and should a rider find he is encountering
very many truly stubborn or unwilling temperaments, he
should examine his own skills and find what it is he is doing
that has made him such a bad horseman.*

Such a passage may alarm someone who is worried about
social architects reasoning from a theory of human nature that
takes the trainers' view, which is that many important character
traits (moral traits) are inborn.† But the moral to take from this
passage is the one about putting your heart into a rogue (which
means risking neck and limb sometimes) rather than take the
far greater risk of being a bad horseman. That is the risk I was
unwilling to take in training Drummer Girl, and that my partner
was unwilling to take. It is also, I think, the risk Drummer
Girl herself was unwilling to take—there is an equine equivalent
of mediocrity, and there are horses who refuse that possibility
at all costs.

As it happens, I never encountered a horse in whose soul
there was no harmony to call on, though I have thought about
giving up more than two or three times. I believe in such
horses, especially when someone like Steinkraus tells me about
them, just as I have come to believe that there are human
mental disorders that are, as it were, the cancer of psychiatry—
meaning that no one knows what to do about them—because
of my respect for the people who tell me about them. I just
never had such a horse in my barn, partially through sheer
luck. But Steinkraus' willingness to acknowledge the existence
of such horses is part of what I want to call his heroism,
especially when I set his passage about rogues next to Will

---

*William Steinkraus, *Riding and Jumping* (New York: Doubleday, 1969),
p. 76, emphasis mine.

†It is not at all clear to me why believing in temperament in humans should
lead automatically to social horrors. Knowing in a general sort of way that there
are genetically determined differences of one sort and another is very different
from knowing what to do about an individual case. Good trainers ride all horses
as though they were potentially world-class; that's the only way you end up
with world-class horses.

James's insistence that "there ain't no horse that can't be rode, and there ain't no man that can't be throwed."

The hero and the heroic horse both know, in one fashion or another, that we must risk death, and that this is the must of logic, as certain would-be suicides know only too well. Ignoring this leads to the moral fatigue and pallid luxury the characters in *Dispatches* found unalterably distasteful. But there is letdown after the glamour of winning—there isn't enough of it to go around, as it were, or enough for any one lifetime, whereas the glamour of being fully alive, angelic in what horsemen call forward desire, creates an impulsion not so easily evaporated on Armistice Day. It is in us not only to risk death but to trample on it, the way some horses trample on rattlesnakes, or else to live in the indoor luxury that makes the thermonuclear button so unhealthily fascinating. A friend of mine wrote to me once that if thermonuclear war comes, it will be because someone got tired; it will not, she said, be an accident, but a deliberate decision to log out. I think she is right, and we mustn't get tired, because if we do, we will end by murdering babies, and it isn't right to murder babies.

To understand safety and genuine risk in relationship to war is part of the prelude to understanding what happens when a horse like Drummer Girl gets sane. The texts of horsemanship begin in a military context—the original one, Xenophon's *Art of Horsemanship*, was a military manual. Drummer Girl's craziness and her sanity have their origins in these texts, in the various traditions of horsemanship that produced her; one may say that she is the artifact of those origins in somewhat the way a physicist might want to say that material objects are the artifacts of gravitational fields. Hence to make mistakes about those origins is to make mistakes about who Drummer Girl is and who we are. Such mistakes are inevitable, but they need not be forever stupid.

Xenophon, writing his philosophical military manual, understood himself to be speaking to the officer or potential officer, and he advises him against breaking his own colts, saying, "How we must break colts, it does not appear to me that I need give any account.... He who assuredly knows as much as myself regarding the management of colts will... give his colt out to be broken."

Xenophon has been so revised that Alois Podhajsky, who was his true heir if anyone was, instituted the practice of accepting only young, unspoiled, green riders and horses for full training at the Spanish Riding School. War in this case has become fully transmuted into art, which my experience with crazy animals leads me sometimes to believe is the only course for either war or Eros if they are not to rob us of ourselves. But Xenophon's advice on all sorts of matters, from the selection of temperament to proper methods of mounting or which exercises are most important in battle, is given with not only the raw exigencies of battle in mind but also with courtesy or courtliness in mind. The rider, for example, when the day's work is completed, is to dismount immediately rather than have the horse carry him back to the stable, be it so much as a mile, for "horses greatly appreciate such courtesies." The last is from a Renaissance translation which adds that, "in the place where the horse is obliged to exercise himself, let him also begin to rest." So it is in the age when horsemanship has become fully figured as an emblem of the state that Xenophon's hints toward elegance are able to enlarge the imagination and teach us to partake to some extent of the stance and movement of mind that produced *The Faerie Queene* and even in our century could produce *National Velvet*.

As a consequence perhaps of our having Xenophon as our original source, a peculiar belief runs among horsemen. I don't know how many, but I was taught it at a respectable riding academy and have heard it a lot since. This is the belief that the movements of high-school dressage, airs above the ground, were developed for use in battle. These difficult and spectacular leaps were supposed to have various functions. The capriole, for example, was a handy way of unhorsing an enemy or kicking him to death. This kind of thinking is part of a general tendency horsemen have to suppose the existence of a Golden Age when all soldiers and horses were capable of the highest demands of the art, but it is very hard to believe there was ever such a practice. The high-school airs require quiet, meticulous, orderly preparation; Shakespeare was nearer the true nature of both the airs and Eros, I think, when in "Venus and Adonis" he uses them as tropes that reveal Adonis' splendor. And Gervase Markham wrote training manuals in which he includes instructions for teaching the airs under the heading "The Pleasure Horse."

I don't mean that the athletic and mental preparation necessary even to begin to teach these airs can't give an already bright, bold horse additional resources that could be used in battle, especially if the horse has a protective, possessive attitude toward the rider, which some horses have, an attitude that would motivate the horse to think well and bravely, as it does good dogs. Indeed it is at least possible—I have no evidence for this beyond the nature of horses—that it was learning that horses are inspired to make extraordinary leaps in battle that led to the formal development of the high school, although, as we will see, Eros is as good as Mars, if not better, as a muse for horses.

The interesting thing is the nature of the mistake, which may be a specifically American mistake, the confusion of creatures with machines, taking too literally the grammar of the horse-as-weapon. Certainly it seems that Americans are more likely than the English or the French to think of the dog as a weapon, which is why the development of good police-dog programs in this country has been so slow.

In this century, in Vienna, at the Spanish Riding School, an ancient cavalry institution, the practice of high-school riding became finally disentangled from the requirements of defense under Colonel Podhajsky, the first director of the school to be wholly relieved of military duties.*

It is hard to imagine a modern dressage trainer advising

---

*Interestingly, the Home Office in London seems to have a similar attitude; in the handbook of 1958 the reader is told that prior military experience in handling dogs does not automatically disqualify a man from being a handler, but it does bar him from the title "Trainer." In both cases, the objection is not to military experience per se, on the grounds that it is, say, too violent, or ruthless, or something, but just to the badness, the sloppiness, of the training.

high-ranking officers not to train their horses too well, as
Xenophon did, on the ground that this made it difficult for the
men to keep up, and that would be a distraction from the duties
of the officer. Nowadays the horses themselves are figures,
tropes, in which nobility and courage can be performed and
contemplated. This is certainly to some extent a function of the
fact that the cavalry is obsolete, but the movement from war
to art began long before anyone could have imagined the obso-
lescence of the war horse. John Astley, for example, a Renaissance trainer who translated and commented on Xenophon,
gives us this version:

> But if a rider teach his horse to go with the bridle loose, to
> carry his neck high, and to arch it from the head onwards,
> he would thus lead him to do everything in which the animal
> himself takes pleasure and pride.
>
> That he does take pleasure in such actions, we see sufficient
> proof; for whenever he approaches other horses, and especially
> when he comes to mares, he rears his neck aloft, bends his
> head gallantly, throws out his legs with nimbleness, and carries
> his tail erect. . . . A rider . . . will thus exhibit his steed as taking
> pride in being ridden. . . .

The result of riding a horse properly, even a war horse,
should be never to have a mount whose energy is the energy
of "anger . . . for horses, when they are annoyed, will assuredly
not use their legs with greater agility and grace." What one
wants are movements that "are at once pleasing and formidable
to contemplate."* This could easily be a sentence from a treatise
applying Aristotle's *Poetics* to horsemanship.

Xenophon insists constantly, as do the other great trainers
for whom I am allowing him to stand in as the exemplary case,
that the objection to cruelty is simply that it doesn't work. The
horse must move with "self-carriage," from his own pride and
pleasure, "for what a horse does under compulsion, as Simo
also observes, he does without understanding, and with no

---

*The disentangling of "Xenophon himself" from the Renaissance versions
of him would take more space than is appropriate here. My point is just that he
was read in this way at this point.

more grace than a dancer would display if a person should whip and spur him during his performance."

Of course, a trainer may be described as *inducing* something or, in a behaviorist vocabulary, as "eliciting behaviors" from the horse. But if that is a possible vocabulary, then "inducing" the horse to elaborate and meditate on his own love of erotic display is to give him an arena within which knowledge and understanding may be developed. If war horses used caprioles to unhorse the enemy, then it was not because the rider performed some series of coercive actions describable within a behavioristic vocabulary, but rather because the courage of art came to command and ennoble the horse.

Observations about the unsettling affair between Venus and Mars have continued for some time in Western thought, and it may seem that I am doing some simpleminded recapping of the claim that "Eros = art = war," taking Tim Page's position when he says that trying to take the glamour out of war is like trying to take the glamour out of sex. But the interpretation of equations suggested by the grammar of the matter depends on looking at what comes after—the differences between Tim Page's possibilities in the world after the war is over and, say, Steinkraus' after he won the Olympic gold medal on Snowbound.

In *Dispatches*, after Vietnam is over and Tim Page is "safe" back in the United States, he is a bored, uneasy man with no philosophically uncontaminated firefights to satisfy him. Steinkraus at this writing is still as passionately involved with riding as ever and is still, in my estimation, a hero, and has this to say, in the second edition of *Riding and Jumping*, the one published after his gold-medal ride:

> I cannot fairly conclude this summary of changes without at least a brief allusion to the very pleasant change of status I myself experienced as a result of the Mexican Olympic Games, for there I finally "caught the brass ring," or became a bride instead of a bridesmaid, or call it what you will. It hasn't exactly changed the way my breeches go on (still one leg at a time), but it was a wonderful experience all the same; and if this book can help someone else to have the thrill of winning

an Olympic medal, or even to experience in some degree the pleasure I've derived from working toward it as a goal, I'll be more than satisfied.*

There was no letdown for Hans Winkler, either. This is a glamour that spreads, an indication of what happens when, instead of trying to take the glamour out of war, you manage to take the war out of glamour. There are examples that are less spectacular as historical events, too.

Drummer Girl's soul, like Tim Page's when he wasn't in a firefight, was several sizes too large for her when the heroic life and its contemplation and preparation weren't available to her. That was what was wrong with the language she had been taught; the heroic had been expunged. Once she was quiet over low fences and during all exercises on the ground, I started asking her to take the genuine risks that are the only opportunities any of us have for a sufficient metaphysics, one our souls will fit.

I had before kept things as safe as possible. Then, one day, we were at a horse show. (Not her first horse show, I should say.) The fences in one class were 4'9" to 5'3", and while this was not a Grand Prix course, those are Grand Prix–size fences. At this level there are no guarantees.

There was a big oxer, followed by a tight, trappy turn to an equally formidable fence. Drummer Girl, on landing after the oxer, pecked and stumbled to her knees. I didn't pull any muscles, but I did, more shamefully, lose the reins, which were floating around the mare's ears. With most horses, I suppose, I would have pulled up then, conceding the ride at that point, but by now Drummer Girl had enlarged herself, her soul was filled out, so instead of grabbing at the reins, I asked, with legs and seat only, that she pull herself together and take the fence. Which she did, rising literally from her knees to jump it, at an angle. There was a wide enough turn to the next fence to enable me to collect my reins and myself, and since Drummer Girl kept her cool, we won that class.

*From the edition of 1969.

Winning was thrilling, but it was that movement, I think, from fall to heroic leap that was the final restoration of a sane glamour to Drummer Girl. And while neither of us went on to the kind of fame Hans Winkler and Halla earned, largely because I am no Hans Winkler, I want to return to Halla's story and say that Winkler is famous for discipline and precision in his riding, for just those qualities a pseudo-Marxism wants to call "coercion." The horses, having read neither Marx nor Foucault, embrace heroic precisions when they can. To say that horses are noble animals is to say, in part, that one of their gods is the infinite god of the details of the surface, plus the god of the studied ignorance of psychological depths and etiological empathies behind the "poor, mistreated horse" stories that had kept Drummer Girl batty.

We, too, are noble animals. I mean that we are born to it, born to the demands of the heroic, of a pleasure earlier than love and nearer to heaven, the pleasure of the heroic approach to knowledge of form. Hence an ethics or a theory of justice or a theory or practice of education that makes no attempt to trick out the syntax* and the semantics of the heroic as a central mode of being human, is not an ethic, whether of animal rights or human ones. Jacques Derrida, writing about Nietzsche's *The*

---

*Calling anything in the exchanges between horses and humans or among horses syntactical as well as semantical is problematic, I know. This is not the place to consider the problem, but I have come to think of syntax as prior to semantics, or of semantics as requiring syntax, since even a gesture or utterance consisting of a single counter, such as "Stop!" is meaningful only in the context of what comes before and after, and there are rules about what can come before and after, rules there is no reason not to call grammatical. There are vast conceptual differences between what people can do with language and what even chimps and gorillas can do, but worrying about preserving the word "syntax" for use in discussions of parsing human sentences will never enable us to understand them. A horse's syntax is limited, from our point of view, by the horsy concept of time, or else the horsy concept of time is limited by the horse's capacity to grasp grammar; and indeed it may be, as John Hollander likes to insist, that our concept of time, which is to say our capacity to have knowledge *that*, as well as knowledge *of*, is founded on the knowledge that we die, a knowledge animals don't have. But all knowledge, all post-lapsarian knowledge, requires meaning-in-time, and thus syntax.

Genealogy of Morals, says, "Degeneration does not let life dwindle away through a regular and continual decline and according to some homogeneous process. Rather, it is touched off by an inversion of values when a hostile and reactive principle actually becomes the active enemy of life. [It] is not a lesser vitality; it is a life principle hostile to life."* In false romances that oppose the quest and the hearth, safety and the heroic, safety becomes in this way "degeneration," but the great literature of the heroic tells us that the quest and the hearth are the same thing, that genuine safety demands the genuinely heroic, or at least that they must be mated if life is to be fecund of meaning. And to ignore this primary aspect of the spirit is certainly not to write a philosophy of consciousness or, it may be, of anything at all.

*"Otobiographies," translated by Avital Ronell, in *The Ear of the Other* (New York: Schocken Books, 1985).

6

Horses in Partnership with Time

Grand Prix riding, which is to say either dressage or jumping competitions in which the demands placed on horse and rider are so extraordinary that it is not hyperbole to say that they are limitless—and in precisely the way that the demands of art are limitless—are places where the logics of art, sport and morality become indistinguishable.* In Grand Prix jumping, the judge's decision rests solely on whether or not the lumber is still there or has been knocked to the ground after you and your horse complete your ride, and on what the automatic timers say about the time the ride took. We call this a sport or a game because neither the horse's nor the rider's intentions enter into public judgments about what has occurred, any more than a batter's intentions enter into the judgment "foul ball" in baseball. You may, for example, have intended to use the course

---

*When I say this, I have in mind Stanley Cavell's discussion, in "A Matter of Meaning It," of the different ways our intentions alter our positions and our responsibilities in the various forms of life that sort themselves out as either art, games or morality. He says:

Games are places where intention does not count, human activities in which intention need not generally be taken into account; because in games *what happens* is described solely in terms set by the game itself, because the consequences one is responsible for are limited by the rules of the game. In morality, tracing an intention limits a man's responsibility; in art it dilates it completely.

simply as preparation for another, more important Grand Prix
event, and so not have put a great deal of pressure on your
horse to jump clean. That is, what you had in mind was
increasing the horse's knowledge and experience, not primarily
winning. But if the fences are all up when you have finished
taking them in the correct order, and your time is the fastest,
you have won. The observer is to a great extent relieved of
the burden of judgment. Dressage events are slightly more
complicated for the judges, since in these, as in gymnastics or
diving, winning is a matter of how well the horse performs a
given movement. There are nonetheless criteria for judging that
are largely independent of the rider's or the horse's intentions.
I don't mean, of course, that such events are commonly won
or lost *accidentally*, only that the logic of judgment does not
demand an account of the competitor's intentions. There are
exceptions to this general rule, as there are in baseball when
the pitcher hits the batter with the ball, but such exceptions
tend mostly to show that even baseball is not epistemologically
perfectible.

In all riding competitions, the presence of the horse imposes
a continuous, unique moral burden. Not all riders are responsive
to this burden, but its presence is revealed in the logic of horse-
show judgments. The regulations and structures of different
horse shows are variously hedged with regulations about cruelty—and whether or not a judge or a committee decides that
a given matter is a case of cruelty will depend, as it does
normally in the rest of the law, on evaluating intentions. In
Grand Prix riding, as it happens, it is rather difficult to compete
both successfully and cruelly, but it is not impossible. It may
happen that while warming up a rider uses a pole studded with
tacks to inspire the horse to jump high and carefully. If the
rider does this knowingly, it is straightforwardly cruel. If,
however, in the confusion of the warm-up ring I have without
noticing taken my horse over such a pole that you or someone
else has brought to put up, or if a trainer, unbeknownst to the
rider, blisters the pasterns of a gaited horse, then the rider isn't
in the same way responsible for what has happened and is
probably not going to be barred from showing. (Though there

might be a reprimand to the rider to be more alert in the future, and, of course, once the rider is aware of such a possibility, the nature of his or her responsibility expands. If my trainer has blistered my horse's pasterns, and I don't fire the trainer or at least get really tough with him or her, then official as well as unofficial judgments of my behavior will change.)

But the burden is deeper than this. It is like the burden of teaching humans, since the nature of riding is such that doing it at all entails meaning to do well by the horse. There are hundreds of children's stories that reveal this; the rider or keeper who sacrifices the horse's mental and physical well-being to some momentary or permanent advantage is shown to be not a bad horseman but rather no horseman at all. The world can make mistakes about this, which is not only why there are such stories but also evidence of what kind of allegory a horse story is. In jumping there are some riders who win fairly continuously for seasons of varying lengths despite rough, ugly methods. These riders' failures to develop their horses' beauty, understanding, nobility and so forth cannot by the logic of the thing enter into a judge's decision whether or not s/he has eyes that are open to genuine beauty as opposed to what is merely thrilling.

The case of the rider who clowns and lurches his or her way around the ring year after year with little or no intelligent thought for the horse is different from the case or the rider who just doesn't happen, for whatever reason, to give a pretty ride; it is the rider's intention to *ride well* that makes the difference. The green rider going around the show ring for the first time on a wise and long-suffering old trooper is not judged *cruel* because of the knocks and bumps both rider and mount encounter and endure. Beyond which, there are riders who are not especially strong, or a bit stiff, or badly built, whose horses nonetheless go beautifully in the full brilliance of their intelligence.

What matters is an understanding of the *horse's* capacity for caring about beauty, precision, perfection of performance. This gives his or her pain meaning and context, so that riding a green horse who knows little or nothing of art close to the point of exhaustion is totally different from, on occasion, asking a mature horse with a strong, developed vision for an effort that will leave him or her for a time weak. It is not that the mature horse becomes desensitized to pain, but that the pain now means something.

To say that pain is meaningful is simply to say that for the creature experiencing the pain there is something that matters more than comfort, for the moment at least; something that is the ground of a certain creature's being, what it cares about, toward which it is oriented, in relationship to which pain isn't quite pain anymore, not anything that matters. This is probably not something anyone can make a judgment about for anyone else. My pain, like my death, belongs to me uniquely most of the time.

My horse, when he is in his stall or lounging about the pasture, has the same relationship to pain that I have when cuddling up with a good murder mystery—comfort and convenience have top priority. Indeed, convenience is so important to horses that in the earlier stages of training one can accomplish a great deal by remembering that, as Podhajsky points out, they will even overcome objections to obeying in order to save themselves inconvenience. Nonetheless, a developed jumper or dressage horse not only doesn't object to the removal of his warm blanket and the substitution of saddle and bridle, s/he actually welcomes these preparations for work. And even horses who are extremely fastidious about stepping on wet, sloppy ground will cheerfully plow through it under tack in the course of performing voltes, serpentines and so on. (Similarly with dogs: my Pit Bull Belle is one of the most comfort-minded dogs in existence and has gone so far as to pull the electric blanket off of my bed and put it on her bed. Furthermore, in the winter, she hates going outside, even to relieve herself, and peers out the door at the weather with extreme gloom. With her tracking harness on, it is different, and when, at the beginning of a track, I ask for the down-stay that lasts for several minutes, she will throw herself down cheerfully in ice-encrusted snow, though not at any other time.)

Some horses are plainly more sensitive to pain than others,

and Drummer Girl was such a horse. If she knocked her legs against something while playing in the pasture, she would actually limp for a while, looking distressed and sorry for herself—and this was not the kind of faking that some horses do so well. Even though she loved jumping, I had to be very careful in the beginning to avoid, as far as I could, her hitting herself on the jump poles and becoming frightened again. Yet once she came to *understand* and actively participate in jumping, she wasn't distracted even by hitting a quite solid fence with a significant thwack—she simply responded by jumping harder and more carefully the next time.

Almost anyone can see what I'm talking about at a Puissance class. In such a class there is a relatively low (around four feet, usually) practice fence, and then a few very formidable obstacles that are raised after each round until there is only one horse remaining who can clear them. The fences may go higher than the mounted rider's head; these are the classes where jumping records are set. Horses do not go about jumping such heights in other situations—at least, I never knew a horse to jump higher than seven feet to get out of a corral, even a horse that can jump higher than that under saddle; a Puissance horse can often be kept quite handily behind fencing that is no more than four feet high. Horses, like people, require meaningful occasions and contexts for intentions that are deep and focused. Seasoned horses develop a very keen sense of the importance or absence of it in a given context.

The riders, while they can tell a horse they know fairly well something about the magnitude of such an occasion ahead of time, can't say, "At around eight-thirty this evening you will have to be psyched up for your best effort," though they *can* say, "Horse! Stirring things are in the air!" So in the Puissance ring there is a practice fence, which is jumped more or less on the way to the fences that count, and which is the last step in the mental preparation of the horse.

One frequent way of using this fence is deliberately to jump it badly. I don't mean to interfere actively with the horse's taking it well, but rather to come in to it casually, sloppily even, not paying much attention, not putting up one's best effort, so that the horse either takes it uncomfortably, off-balance, or actually hits it. If it is the right horse, the right rider and the right training, this throws the horse powerfully onto his own mental resources, and you can see good jumpers, after the practice fence, prick up their ears, look around for the next fence and instead of trying to pull away from it, pull eagerly toward it. You can see even an awesome moment of decision in horses coming down toward really big fences.

First there is the discovery, the moment of "My God! That is a big sucker," and then the deliberate gathering, in the horse, of all of her power, all forward desire. (Or, unfortunately, sometimes the opposite, a desperate squirreling around, seeking any avenue of escape.)

There are various ways to talk about what could possibly motivate a horse, or any animal, to such an effort. Fear certainly does not do it. Courage, joy, exaltation are more like it, but beyond that horses have, some of the time, a strong sense of artistry. This is something very specific. It is not merely crafts-manship, although being able to do a difficult thing well is, of course, a powerful motivator for man or beast. When I say artistry, I mean that the movements of a developed horse, the figures and leaps, mean something, and an artistic horse is one who is capable of wanting to mean the movements and the jump perfectly.

The jump, like the complicated movements of dressage, is an imitation of nature, especially of various movements that horses perform for the sake of sexual display or in the course of exercising claim rights—a stallion's claiming of a herd or a mare's claiming of a foal or of leadership status. In nature, in the horse's first inheritance of these gestures, they have particular meanings, such as "Wait your turn!" or "Watch out!" or "Look at me! I love you!" or "Wait. Be still. Something wicked this way comes." But the movements of dressage and formal jump-ing, properly performed, don't mean "Look at me, I love you" or any of the others. They mean the natural movements them-selves. This is why the language of analysis and criticism of riding at the highest levels is the language of art criticism. Podhajsky, attending the Olympic games, recorded detailed observations of the rides, which became the book *The Art of Dressage*. He says, speaking, as always, after Xenophon, "Anything forced or misunderstood can never be beautiful," and later in that same chapter:

> ... nature can exist without art, but not art without nature. Consequently, the well-trained dressage horse should perform the natural paces with perfection. Any defects in these movements cannot be made up for by some other spectacular exercises. Riders or judges who allow themselves to be dazzled by such striking movements betray the true art of riding.

Such betrayals are not only common, they have become activities in themselves, and there are competitions in which special effects have entirely replaced genuine art. In the true art, Podhajsky insists, haste in the movements is always a fault, as is unhappiness on the face of the horse. Guiding phrases include: "purity of the paces," "*harmony*, lightness in all movements," "the impression of complete confidence," "the horse's *concentration* upon his rider becomes obvious. . . . " Riders are criticized for failures of *cadence* and *tempo*, or are praised for what is "*fluent* and precisely performed" and especially for what is *expressive*. Of a ride that does not achieve full entry over what he calls "the threshold of art," he says, "without brilliance. . . hesitant transition. . . performance of a tolerably obedient horse with little charm and suppleness."

Podhajsky died before *The Art of Dressage* could be published, so it concludes with an "In Memoriam" by Berthold Spangenberg, who thanks his great master for teaching "the gentle, the difficult art of riding." And indeed it became a gentle art under Podhajsky's influence, but gentle because so difficult, or gentle because genuine, since in a genuine art there is the amassment and expression of true power as opposed to mere force. Which brings me finally in contact with the central question, which will be hard to talk about because it is difficult to talk about the horse without talking about the rider and vice versa (although it can be quite hard to see why from the ground). If this is an art for the horse, then the horse must intend it, and what does the horse intend, what can the horse possibly mean by it? And does—can—the horse mean by it anything like what the rider means, or can mean?

The movements themselves are not literal; they don't mean what they say. The horse would be as disconcerted if anyone, equine or human, were to respond to, say, a capriole as a literal threat or appeal, as a composer would be to such a response to passages of music that are expressive of rage. But if the movements are not literal, the horses nevertheless mean something by them, and there are a few horses, very great horses, who, like very great artists, have the capacity to accept full responsibility for meaning what they "say"—do—and this is the kind of meaning that is always entailed in art. In other words, I am claiming that a great horse, in Cavell's words, "is responsible for everything that happens in his work—and not just in the sense that it is done, but in the sense that it is meant."

This does not yet explain what the horse means by it, but it does begin to suggest one aspect of the rider's role, the rider's responsibility. Cavell goes on to say:

> It is a terrible responsibility. . . . But it is all the more terrible, when it is shouldered, not to appreciate it, to refuse to understand something meant so well. . . . In art [the right to question the artist] has to be earned, through the talent of understanding, the skill of commitment, and truthfulness to one's response—the ways the artist earned his initial right to our attention. If we have earned the right to question it, the object itself will answer: otherwise not. There is poetic justice.

So a rider who is a true rider and no mere keeper of horses is someone who continuously earns the right to question the horse and the horse's performance, and it is the horse's performance that answers the rider's questioning. Thus the very fact, the very possibility, of Grand Prix riding, both jumping and dressage, is our discovery in the horse of a capacity for meaning a movement or a series of movements artistically.

When the threshold of art has been crossed, then the wonderful obedience and supple submission of the horse, the joy of the horse's submission, are like the intensely accurate responsiveness of a great performer to a good audience, another case of the collapse of command and obedience into a single supple relation. It is, as Podhajsky says, as though the rider thinks and the horse executes the thought, without mediation or any sort of cuing; but it is also the other around way on the back of a great horse—it is as though the horse thinks and the rider creates, or becomes, a space and direction for the execution of the horse's thoughts. The rider is the person who shoulders the burden of knowing through "the talent of understanding and the skill of commitment" what the horse means.

But, as I have said, there are differences between the horse's concept of time and ours. It is, of course, the rider and not the horse who sits in the study working out training schedules and filling in entry forms to be mailed to the show committee, marking dates on a calendar. Horses do not have what we call the tenses of verbs, so they don't talk or think about that, and that is why the Puissance rider, for all of the wonderful things s/he can say to the horse, can't say, "Be ready to do your very best tonight at eight o'clock." The concepts of time that enable us to make appointments and leave notes are not in the grammar of the horse's world, so we cannot share that form of life with them. And our concept of a rehearsal is very much a concept of ordering time in a particular way, by means of a particular grammar.

There is a novel called *Nightmare*, by Piers Anthony, which is told from the point of view of a mare whose task is to bring instructively awful dreams to human beings. One of these human beings is a woman who, herself rapidly approaching death, asks the mare if her kind, horses, are immortal. The mare, for whom the question is a new one, thinks about it and comes up with the clumsy but strangely apt answer, "Yes... until we die, that is."

But some horses (and this, alas, is part of the talent in them) are capable of responding to the knowledge art creates of what it feels like when there is complete congruence between the soul and the moment (that congruence Wittgenstein indicated when, at the end of the *Tractatus Logico-Philosophicus*, he says, "He who lives in the present lives in eternity"*) with a general anxiety that work should continue. When left to mooch serenely around the pasture, some of these horses become, not peaceful and lazy, but rather depressed if work for some reason ceases. Such a state of mind does not, of course, entail the full grammar of angst or melancholy in human beings, but the taint of mortality is on the horse, and it takes more than mere comfort for his or her spine to be restored to a feeling of congruence with the landscape. I don't mean that in such a case any horse ever learns that death is inevitable, any more than any horse ever learns that the National Horse Show at Madison Square Garden is on the twenty-seventh of next month, however much s/he learns about the significance for a given moment of activities before, during and after horse-show activities and seasons. We cannot, thank heaven, teach horses the tenses of English verbs, which means that we cannot teach them that they die, or that we do, any more than you can tell your dog not to worry, that you will be back from the store in ten minutes, or back from the Holy Land in ten years. But we can nonetheless interfere with, disturb their sense of time, teaching new modes of anticipation as well as new modes of loss.

The book *Nightmare* provides a kind of allegory of this. The heroine becomes "mortal by day," and this means that, during the day, she cannot escape being ridden by the extremely sinister figure called Horseman. That vulnerability is a nice emblem of what horses can learn from us—not the grammar of mortality, not a knowledge of their own, but rather a participation in ours.

If what I have said so far were all there is to it, it wouldn't be hard to come to the conclusion that training horses is morally indefensible, but this isn't all there is to it, because horses have their own grammar of time. They can't say anything that requires past, present or future tense, but that doesn't mean that without us they live in eternity, in the present tense only. Their concept of time might be expressed by saying that the

*"*Denn lebt er ewig, der in der Gegenwart lebt.*"

names of their tenses are "not yet, here and gone." You can't make appointments with such tenses, but you can remember, and you can anticipate the future with no little anxiety. That is to say, horses do have some sensitivity to the knowledge of death, and it makes them nervous, just as it makes us nervous.

That knowledge is what they are relieved of, just as their riders are, in the tremendous concentration of horsemanship at the highest levels. This is why we are forgiven for riding them, especially in competitions, for distracting and scaring them with brass bands at football stadiums, spooking them by placing garlands of roses around their sweating necks and surrounding them with photographers, neon beer signs and journalists who profanely scribble figures on note pads while the horses are jumping their hearts out. And nothing short of the tremendous artistic task of training them in such a fashion that they can be released from time could ever justify our interfering with their greater serenity, our imposing our stories and our deathly arithmetics on their coherent landscapes.

What they mean by their artistry, then, is just this, which one could call the release from time, but which could also be understood as what happens when a horse becomes time's lover or time's partner, moving with time instead of as time's slave.

7

Calling Animals by Name

*And Adam gave names to all cattle, and to the fowl of the
air, and to every beast of the field. . . .*

Genesis 2:20

In the course of restoring Drummer Girl to herself, I obedience-
trained her, and in the course of doing this work with me, she
learned what her name was. In fact, although there are often
problems even for humans about learning their names, about
knowing what one's own or another's name is (as when I don't
know whether to call you Freddie or Professor Jones), for us
naming the animals is the original emblem of animal responsiveness to and interest in humans, in Genesis, our first text.
An apocryphal expansion of the verse that forms the epigraph
to this chapter says that not only did Adam name the animals
but the moment he did, each recognized his or her name; the
cow now knew she was Cow and came when called by name,
and so it was like this, as John Hollander describes it:

> Every burrower, each flier,
>   Came for the name he had to give:
> Gay, first work, ever to be prior,
>   Not yet sunk to primitive.

Now it is the case, sadly, that many horses go through their whole lives without even knowing that they have a name, and this misleads some logicians into believing that they can't have names, and therefore can't have the mental faculties that go with knowing one's name. But in fact many horses learn their names, either informally around the barn or stable, just as most humans learn their names, or through formal obedience work of the sort I did with Salty and Drummer Girl.

I would like to take a little time here to consider the general implications of naming and acknowledging naming. I see us—meaning anyone possessed of that particular sort of literacy that makes him/her want to write and read books like this one—as not being in the enviable position Adam was when he named the animals—"not yet sunk to primitive." I don't mean that we are primitive in our consciousness but rather that we have gone on to a further distancing. We did this when we learned to write and thus to add to the possibilities of consciousness conceptions made possible by typography of various sorts. One example of this is the advance in mathematical thinking when numerals were devised and replaced the prose descriptions of arithmetic. It was typography that eventually made statistics possible and all of the errors as well as the epiphanies of statistical thinking.

Typography has also made possible further gaps between us and animals, because we have become able to give them labels without ever calling them by name. The registered names of most horses and dogs are primary examples. Champion Redheath Nimble Gunner, C.D., C.D.X., U.D., for example, is not a name but something halfway between labels (of the sort found on packing lists or in livestock inventories) and titles—not titles such as Sir, Madam or Your Highness, but titles like the titles of books. Such names are bookkeeping.

It is only when I am saying, "Gunner, Come!" that the dog has a name. His name becomes larger when we proceed to "Gunner, Fetch!" and eventually when he and his name become near enough to being the same size, he is as close to having a proper name as anyone ever gets. When Drummer Girl learned

her name, one of the things it meant was that she became able
to fit into her name properly; when I said in her story that
"her soul was several sizes too large for her," I could as
accurately have said that her soul was several sizes too large for
the truncated version of a "name" she had so far had, not a
name she could answer to. Without a name and someone to
call her by name, she couldn't enter the moral life.

There are other things at stake. I knew a woman named
Shelley Mason, who took a job running an animal shelter in a
small desert town. She didn't do this because she thought that
the activity of merely housing dogs and feeding them was an
especially meaningful activity (especially as one of her duties
was destroying unwanted animals) but because she understood
the importance of training as a way of increasing the number
of animals who were wanted and who would not be abandoned,
thus reducing the piles of corpses. She figured that from that
small shelter she could insist that anyone who adopted a dog
learn at least the rudiments of training a dog to heel and sit
before they were allowed to take the dog home, thereby of
necessity naming the dog. And she usually had several dogs
from the shelter at her house, teaching them more advanced
work in order to increase their chances of placement.

One day when I was visiting her, she gestured at the dogs,
most of them doomed, in the runs at the shelter and said,
"Goddamit! Most of them wouldn't be here if only they knew
their names!"

The grammar of the world we imagine when we call creatures by name is not the grammar of the world in which they
have no names, is not the same form of life. But our grammar,
or maybe I mean punctuation or typography, has given us the
possibility of attenuations of naming, of names that are not
invocative. Consider for example that

I am involved with a dog

does not indicate a world as fully as when we say

I am involved with a dog called "Annie"

or

I am involved with "Annie."

The last example gives the feel of a more committed and thoughtful relationship than the first two do, but it is still a disturbing (to me) convention of English punctuation to put what philosophers call scare quotes around animal names, to indicate that these aren't real names, in the way Vicki Hearne is, and even many animal lovers conventionally use the pronoun "it" rather than "he" or "she" to refer to an animal. I find this to be extraordinarily weird, evidence of the superstitions that control the institutionalization of thought. It is as weird, to me, as these examples:

> I am married to "Robert."
> Pass the butter, "Robert."
> Kiss me, "Robert."
> I wish "Robert" would return.

When I asked my husband, whose name in fact is not "Robert," but Robert, to look at those sentences, he reported feeling a slight jolt of uneasiness, as though what had been a name for a person—his person—had suddenly become something like a label, and the uneasiness—the dis-ease—is the uneasiness of someone the labeler won't and can't talk to.

Obedience-training horses creates a logic that demands not only the use of a call name, since the imperatives demand it, especially for the command "Dobbin, Come!" but also the removal of the quotes from the name, the making of the name into a real name rather than a label for a piece of property, which is what most racehorses' names are.

Which leads me to my final small point about the disciplines of naming, one of which is horse training. I believe that the disciplines come to us in the form they do because deep in human beings is the impulse to perform Adam's task, to name animals and people as well, and to name them in such a way that the grammar is flexible enough to do at least two things. One is to make names that give the soul room for expansion. My talk of the change from utterances such as "Belle, Sit!" to "Belle, Go find!" is an example of names projecting the creature named into more glorious contexts. Our awareness of the

importance of this is indicated, at least partially, by the fact
that we have occasions to say, "Well, Rosemary has really made
a name for herself."

But I think our impulse is also conservative, an impulse to
return to Adam's divine condition. I can't imagine how we
would do that, or what it would be like, but linguistic anthro-
pology has found out some things about illiterate peoples that
suggest at least names that really call, language that is genuinely
invocative and uncontaminated by writing and thus by the
concept of names as labels rather than genuine invocations.

I once, for example, heard a linguist talking about the days
when the interest in learning and especially recording illiterate
languages revealed some surprises. One of his stories was about
an eager linguist in some culturally remote corner trying to
elicit from a peasant the nominative form of "cow" in the
peasant's language.

The linguist met with frustrations. When he asked, "What
do you call that animal?" pointing to the peasant's cow, he got,
instead of the nominative of "cow," the vocative of "Bossie."
When he tried again, asking, "Well, what do you call your
neighbor's animal that moos and gives milk?" the peasant
replied, "Why should I call my neighbor's animal?"

Since I am a creature born to writing, my horses are not
born to their names but to their labels, and care and discipline
are required. The dog trainer's knowledge of genuine names—
"call name," in fact, is the technical term for a true name—is
one of the reasons true trainers say, as I reported in my discussion of Salty, "Joe, Sit!" less frequently than most people
and to fewer dogs. They know what the peasant in the linguist's
story knew—there has to be a reason for a name or else there
is no name.

I am not arguing against advances in culture, only pointing
out that it is paradoxically the case that some advances create
the need for other advances that will take us back to what we
call the primitive, even if not all the way back to paradise, to
that region of consciousness in which naming is "Gay, first
work, ever to be prior, not yet sunk to primitive." But no

advance will enable me to call Drummer Girl with anything less than her name, which is why obedience training is centrally a sacred and poetic rather than a philosophical or scientific discipline.

8

The Sound of Kindness

*...it is fatal for anyone who writes to think of their sex.
...It is fatal for a woman to lay the least stress on any
grievance; to plead even with justice any cause.... And fatal
is no figure of speech.*

VIRGINIA WOOLF, A Room of One's Own

There are people who, when they can, inhabit a moral world
that I can best indicate the nature of by taking a phrase from
Moravscik's discussion of *The Illiad*. "Here, kindness is not an
excellence." It is not that such people are not saps about animal,
child and even adult welfare. Such people exert themselves
endlessly helping other people with their dogs and seeing that
children get protection and justice, activities that sometimes
entail risks. But they are also the people who speak contemp-
tuously of "affection training," "kindness" and "the dog who
needs only understanding."

This is a Nietzschean maneuver; it is not kindness itself that
is being refused but rather the word "kind," because the word
has become contaminated, as important words tend to. They
are giving the word away to the kindly people who think
homosexuals, Pit Bulls and I don't know what else would be
better off out of their misery. It's a good word, however, and
I want it back, so I want to tell some small tales about what I
as a trainer see as the soul's eternal battle with unreal forms of

piety and righteousness. I wish I could claim never to have participated in the truncated ways of being the "baddies" the characters in my stories display, but I can't.

Grandma and Grandpa and a herd of cousins are sitting down to Thanksgiving dinner. Freddie has been training his new dog, a good, tough-hearted little brindle Cairn Terrier named Sue. Since Grandma tends to have palpitations and stomach trouble when confronted with Freddie's cruelty to the poor sweet dog, who obviously doesn't want to do all of that regimented stuff, Freddie has so far been discreet about how he works Sue.

Now, however, it is time for Sue to have more experience in distracting situations like large family gatherings. That much in the way of noise and tempting smells around which to have Sue practice her civility isn't that easy to come by, so the dog trainer in Freddie triumphs over the diplomat. The chances of his having to give offense by correcting Sue are remote, Freddie figures, since all he is going to demand of her is that she stay quietly in the general vicinity of his place at the table. Grandma, however, gets the idea of feeding tidbits to Sue from the table and of chirping about it so that everyone present knows how cute, generous and kindhearted Grandma is.

Freddie hasn't had occasion to think over the feeding-the-dog-from-the-table issue, mostly because it wouldn't occur to him to do so, but now he is revolted and enraged at the sight of his gallant little dog being encouraged to beg and flirt.

"Don't feed her!" roars Fred.

"Why not?" says Grandma, her tone a cross between whine, pout and simper.

"Because I said so!" shouts Fred, forgetting entirely about keeping family peace.

Grandma is upset, hurt and physically ill; she takes to her bed. Mother, who dislikes Grandma as much as Freddie does, might or might not at this point reprimand Freddie. Freddie, having achieved his training goal, might or might not be mature enough to let it go at that and might or might not be alert to the shoddy, dishonest training methods that have made

Grandma what she is. The worst thing that could have happened would have been for Freddie to allow himself to be confused enough to permit the degradation of his dog.

Perhaps I should write more forgivingly of Grandma, whose discounting of both Fred and his dog seems more misguided than strictly speaking cruel—*tout comprendre c'est tout pardonner*. But there is the Talmudic aphorism as well: "He who is kind to the cruel will end by being cruel to the kind." If this were psychological realism I would be obliged in my narrative to reveal Grandma's pain, sincerity or whatever. But this isn't realism, it's dog training. There are good guys and bad guys, and it is easy to figure out which is which simply by observing their dogs. There are those who have lost their souls and those who haven't. There are minds like Grandma's, which from the trainer's point of view run to a criminal cunning it is no kindness to encourage. The poet, or the priest, may be under limited circumstances touched with the sort of divine inspiration that makes it possible to paint a workable picture of universal forgiveness. But this is mostly not the fate of the trainer, whose gods are Apollo and Diana.

I used to wonder why trainers existed; giving an animal to someone else to train for you seemed to me as queer a business as giving your spouse to a surrogate husband or wife for training in the conversation that is marriage. It's not clear to me what you would want a spouse *for* under such circumstances.

I drove out to a neighboring town to pick up a horse I was to have in training. This was an Arab, a valuable young breeding stallion. Handsome he was, with many splendid points to give to his foals. But he hadn't done much breeding lately because he had turned killer; no one could handle him. He was housed behind six feet of solid metal piping, and no one had been in his corral for a year, except for the farrier, who brought the vet with him when he came so that Ibn could be tranquilized from over the fence. Hay was cautiously tossed into his corral twice a day; that was the extent of his contact with people.

His owner trotted along beside me as we walked from the house up to Ibn's corral, chattering about how the real trouble with the horse was that there wasn't anybody who understood him and realized how sensitive and tenderhearted he was. As we drew even with the corral, I unwisely turned my head away from the horse to make some sort of response to the chatter beside me. Ibn's head slipped like lightning over the fence, and he bit me viciously in the arm, penetrating through a thick coat and sweater to tear muscle and draw blood. His owner's patter didn't lose cadence for a second, though the theme modulated into a coy one about naughty horses, and she shook a flawlessly manicured forefinger at Ibn, who regarded the finger down the length of his nose in cross-eyed speculation.

I trained the horse, and he was fine, a juvenile delinquent rather than a criminal proper. He did attack me at first from time to time, with teeth and hoofs, and I got quite rough in response to this, having more than one occasion to strike him where the sun never shines or anyplace else that came handy. The roughness of my handling was as mysteriously invisible to his gentle owner as the bite on my arm had been. Suddenly I understood the essential, the metaphysical function of animal trainers. There has to be a separate class to get bitten, kicked and stomped on so that the humane and schizoid patter can continue without interruption. Like policemen, trainers exist on the frontiers of social order, as a kind of windbreak.

I am having coffee with a friend. She tells me about a dog her family had when she was a child. The dog had taken to killing the neighbor's rabbits, and the efforts of her brothers and her father to stop the dog's marauding brought her to a grief-stricken realization of the wrongness of thwarting a natural impulse in an animal, such as hunting. I wondered what certain hunting-dog trainers I have known would think of interpreting tearing up rabbit hutches as an instance of "hunting," but I only asked, "What happened to the dog?"

"Oh, of course, we had to have him destroyed. We couldn't have him killing rabbits."

* * *

Mr. and Mrs. Beacon have known me for a decade and a half. During that time they, especially Mrs. Beacon, have talked to me about how much like the Gestapo my training activities are. I have wondered from time to time about the flabbergasting rudeness of this, but I haven't had the wit or stamina to counter the Beacons' theories. They have a dog. Mrs. Beacon likes to take him on walks. I urged them to take him to a training class for a while, but Mrs. Beacon didn't want his freedom restricted by a leash.

Recently the dog chased down and killed a cat. Now Mr. and Mrs. Beacon want my professional advice. In particular, they want me to tell them how to use a radio-controlled electric collar. Belatedly, my rage rises.

"You want *me* to tell you how to apply twelve volts directly to the throat of an animal when you don't even care enough about him to teach him basic manners?"

Mr. Beacon is better than I am at keeping his cool and says wistfully and with sweet reasonableness, "But Amanda really misses her walks with the dog."

I leave abruptly and uncivilly. Mr. Beacon follows me to the door, and as I walk down the driveway I hear his cultured, amused voice saying, "Forgive, forgive!"

The incident will make amusing conversation about the simpleminded dog trainer—dedicated, perhaps, but with little social or philosophical cool.

Or so I fervently hope. A more complicated response on their part, one that contains the dawning of some confused awareness of the personhood of their dog, will lead to phone calls with requests for further advice, advice that will never be understood, much less taken. I don't want to talk to them about their dog. I have too much to be silent about.

The chairman of the Department of Social Psychology distrusts authority. He is, he explains to me, a kind man, a supporter of feminism, an opponent of racism. He is especially worried about anything that looks like Nazi Germany and is such an expert on things that look like Nazi Germany that there is hardly ever a need, when he is holding forth, for a second opinion. One day at lunch I offer the thought that maybe the worst thing we, as heirs to the knowledge of the Third Reich in various ways, have to deal with is our tendency to congratulate ourselves on not being members of the SS, as though to have escaped that fate were to have succeeded once and for all morally, to be in no further danger of being deranked from the branch of humanity. He replies that my way of talking is dangerous nonsense, to be expected perhaps of an animal trainer who might be unable to see the dangers of obedience. Naturally, when my friend Sheila, who teaches in his department, begins training her dog because she wants to have him with her for protection when she walks to and from campus to work, especially at night, the pious chairman is appalled.

As well he might be. Slender, athletic Sheila stops in front of the door to her office with her large crossbred Shepherd. Before getting her keys out she commands, loudly and sharply, "Sean, Sit!" Sean refuses to sit. Sheila corrects him emphatically. He sits. She enters her office.

Later in the day, the chairman affably invites her into his office and suggests that she shouldn't wear pants to campus. (Sheila is very pretty.)

A bit later in this story, the sorrowful chairman informs Sheila that one of her students has complained about intimidation, in particular about the intimidation of an instructor's having an "attack-trained" dog in her office. Sheila says that Sean isn't attack-trained, so he adds various remarks about how cruel dog training is, how making dogs sit makes them vicious, with embellishments from his private version of the Holocaust.

The same student later threatens another teacher with a knife because he had read a newspaper article in which the state legislature was reported as having voted to remove the law that made homosexuality an offense, and the student is sure that the teacher is behind all attempts to allow such monstrosities as unpunished homosexuality to exist in the free world.

In the meantime Sheila defiantly trains her dog and brings him to school and wears her jeans. This causes much comment. The comment has to do with certain confusions: Sheila is plainly a victim of sexism, on the one hand, and therefore a deserving person, but on the other hand she thinks repressively, makes her dog sit on command. Hence Sheila and the people at the Rape Crisis Center are at loggerheads, because Sheila wants to introduce changes in the campus regulations about dogs.

She is moved to do this because the chairman has invoked the little-observed No Dogs in Buildings rule in an attempt to bring Sheila's cruel training activities to a halt. He has reported her to the campus police. I speak to him, suggesting that reporting a faculty member to the police is surely an extreme measure, uncalled for in a situation in which the quarrels are philosophical. He spreads his hands helplessly, this enemy of fascism, and says that rules are rules. Sheila is trying to get the people in Rape Crisis to remove the No Dogs in Buildings rule and replace it with one that would permit responsible dog owners to have their dogs with them, *on leash*. She has drawn up a list of considerations that make "responsible dog owner" about as unambiguous as things ever get, and points out that there is in any event already a leash law in the town in which the campus is located. She gets no support: the idea of leashes interfering with the dogs' freedom repels the free spirits at the Rape Crisis Center.

This is, incidentally, a campus that prides itself on being the campus for the handicapped. One day a quadriplegic student in a wheelchair is held at bay for an hour by a large dog who didn't like the sound of the motor on his wheelchair. Wheelchair students cannot sit outside in the Commons on fine days and eat lunch, because loose dogs jump into their laps and take their sandwiches. A blind student is mauled by a loose dog who attacks his Seeing Eye dog. And at the end of each school year roving packs of dogs are formed from the free dogs left behind by the liberal-minded students.

Back to Sheila and the chairman. Within a month there are three rapes—in the library, in the women's rest rooms. Sheila has to walk a mile and a half to and from campus on ill-lit roads and isn't, I should say, an especially tolerant sort. When she explains to the chairman about the rapes and about wanting to feel safe walking to and from campus, he says that she had best do her work in the daytime. She, with no regard for his tender political feelings, tells him when and where he can do *his* work.

Sheila is not rehired. I don't know the quality of her work and can't comment on it. Some months later, at a party, the chairman tells me about his sabbatical plans. They include taking his *dear* little dog to the pound, as his scheduled tour includes England, where it is illegal to bring in dogs without a six months' quarantine. The scotch has been flowing; he weeps a bitter tear.

There is a television on at the party, giving a report about some local quarrels over the best methods of killing dogs. A neighboring county is abandoning the method of decompression in favor of injecting sodium pentobarbital. The disputes have to do, not with expense, but with the fact that injecting sodium pentobarbital requires a veterinarian's skill, and the activity is bad for the peace of mind of the veterinarians.

I meet with my friend Dick Koehler to hear about his plans for a training program for dogs for the handicapped—the deaf, the quadriplegic, the aged, the mentally retarded and so on. He hunches over outlines of training schedules, his eyes electric with the vision of personhood and power such dogs will give to people. The dogs, by pulling wheelchairs, ringing doorbells, opening restaurant doors, acting as portable railings and so on, will be wonderfully convenient, but this is not a technological vision; it is rather a vision of the possibility of wholeness of soul. It is, for instance, easy to condescend to someone in a wheelchair, much harder to condescend to eighty-five pounds of agile, muscular knowledge. The principal value Dick sees in the dogs is not psychological but spiritual.

Later I listen to an argument against the use of Seeing Eye dogs. The blind, goes the objection, become too emotionally

dependent on their dogs. My friend Nellie, herself blind and a
dog trainer, says, "Look! Most of the blind are emotional
klutzes. Are *you* willing to put up with them twenty-four hours
a day?"

I try to tell a moral philosopher about Dick's dogs-for-the-
handicapped project. He is philosophically offended at the idea
of the very strict and thorough training involved; it is axiomatic
for him that it is wrong to interfere with an animal's natural
impulses in order to have the animal serve human beings. We
are having this discussion in the university cafeteria. Outside
of the window, assured and radiant, a Guide Dog and her
mistress go by on their way to class.

It is graduation night for one of Dick Koehler's novice obedience
classes. In these classes, graduation night is a competition, run
exactly like a licensed dog show. This is the first time I have
seen this particular group of dogs and handlers.

There is one chap with a Great Dane. He is a man in his
fifties who stands out in the group of people milling around
with their dogs because he is working about as hard as anyone
can at *not* correcting his dog. The sight of him inspires instinctive revulsion in me, as does the sight of anyone who encourages
a dog to misbehave, especially if they, while doing so, look
around smiling genially to see if anyone is admiring the display
of "love."

The dog, although he is out at the elbow and lamentably
straight in the stifle, has a deep and forgiving heart in him:
he performs badly in the ring but is plainly doing his damndest
to figure out what would count as upright behavior. Proof
of this is in the fact that both the man and his wife are
unharmed, and the dog has not split for the next county.
Before the dog goes in the ring, there is a husband-and-wife
spat, with the wife going on about how harsh and cruel Dick
Koehler is.

When the dog comes out of the ring, with a score that
translates into a grade of roughly F+, the wife rushes at him,
breathing petulance and resentment and slapping at him in a

weak, ugly way with her hand. Then she takes the leash and occupies herself for the remainder of the evening using it to hit at the dog in random intervals of mutter.

It must be understood that this is a *big* dog. When he and his handler approach the judges' table, where I am seated, his nose is on a level with my hat. Stop for a minute and think about this. Imagine looking up from wherever you are seated reading this (and unless you are Wilt Chamberlain you will in all likelihood have to look *up*) into the steady eyes of a Great Dane.

There can be disputes about how to read the dog's behavior, of course. Perhaps a lover of tigers would find in it an occasion for contempt. Why isn't the dog standing up for himself? Why does he meekly endure the humiliation? One modern answer is that the dog is stupid, geared by both instinct and coercive ideologies to submission. But this is a hard answer to sustain belief in when you are actually within sniffing distance of 170 pounds of forbearance, which is why the traditional explanation is that dogs are noble.

My husband is with me, watching the exercises. He says of the man with the Great Dane, "The only part of the dog that exists for him is his eyes. The rest is smoke!"

My husband and I meet Dick Koehler after graduation ceremonies for coffee and sandwiches. Dick starts telling me a story about a man named Clarence and his dog Duke, a Doberman. When Dick was in Colorado, Clarence trained Duke under him. He did a good job, too, apparently, because the two of them started cleaning up, as we say, at the matches, time after time getting near-perfect scores.

The Denver Dog Club barred him from matches. He could, they decided, win only one trophy with each dog, because otherwise it wasn't fair to the others.

My husband and I are speechless. We, too, say, "It isn't fair!" Dick shrugs and says that Denver is a cow town, and that's the kind of justice you get in cow towns. I am pleased by this notion of utilitarianism as cow-town justice, and not because I have anything against Denver or cow towns. I say something about utilitarianism and how Dick's story shows what justice looks like when you take out the notion of the Good, the True and the Beautiful.

Dick, however, is in a mood, and says to me, "You don't believe in justice, do you? Nobody with any brains cries, 'It isn't fair!' after they turn twenty-one."

But I cannot remember, in the last twenty years, being in Dick Koehler's company for more than five minutes without finding him engaged in ensuring that justice be served—that is, ensuring that good handling and honest-heartedness get recognized, so I point this out to Dick. The evidence, I claim, is that not only does Dick Koehler believe in justice but serving her is the foundation of his life.

Dick dismisses such sophistries: "Well, that's like poetry, a fairy tale, a never-never land."

Poetry equals dog training, then, because both are clearings, off in some corner of the landscape, somewhere set aside in the service of justice. This reminds me of Spenser, so I start telling Dick about Artegal, the Knight of Justice, and his squire Talus, and it pleases me, not for the first time, to think of the tale of Talus as the first tale of the police dog, or as a prototype of such tales, of beings who can act coherently when the knight can't in circumstances in which the shifting fog of history has obscured the Good.

I forgot to say that Clarence was black. I don't know how this changes or adds to the story.

The story about the man, the Great Dane, the wife and the leash is a story that strikes a dog trainer—I mean a real dog trainer—nearly dumb with its force as an allegory of cruelty. So it matters at this point that I say that even the wife's slapping at the dog with the leash didn't cause enough physical pain to make the dog so much as blink—a gnat pounding on an elephant with a protein molecule would cause more physical pain.

Dick doesn't swear very often, but he swore watching the couple and the Great Dane, saying, "It's a waste, a goddam cruel waste!" There exist various forms of outrage in our species, and most of them are, indeed, specious, irrelevant mechanisms of grievance. But Dick's outrage was an outrage of the soul, which is distinguished from other kinds not only by its tremendous power but also by its accuracy. It is accurate the way lightning is accurate, and indifferent to accusation the way Mount Everest is indifferent to accusation.

Try to understand this. Look in your imagination at the noble head of a Great Dane, the resentment of the owners, and understand what is being wasted. The dog, like any creature possessed of a soul, is immortal, until he dies, that is. And what is being wasted is his immortality, and infinity, too, because mishandling a dog is that sort of offense.

There is such a thing as the end of infinity. One of the traditional words for the end of infinity is death. And the splendor of dogs and handlers is a splendor it is death to hide.

Literally. A trip to your local pound or humane society should make this clear.

There are forces in the world these days battling for equal rights. As the story of Clarence and his Doberman makes clear, one of the biggest impediments to equal rights is the way some people have of writing, dancing, politicking, pole-vaulting and so on more effectively than other people. There's a problem about ensuring that the virtuous, deserving and incompetent poets don't get their voices drowned out by unfeeling people like Shakespeare.

This is a problem. I think of Virginia Woolf's tale about Shakespeare's sister. In her story, the sister was as gifted as her brother and, like her brother, took herself off to London and knocked on the doors of the theater. She was scorned and mocked by everyone, except by the manager, who took pity on her so that she found herself with child by him. And killed herself, and Woolf says, "Who shall measure the heat and violence of the poet's heart when caught and tangled in a woman's body?"

There is also an equal-rights effort going on these days in the dog world. The debate is presided over by the American Kennel Club from its headquarters in Manhattan. The situation is like this: Current rules for the competitions called Open and Utility, in which dogs earn the titles Companion Dog Excellent and Utility Dog, require among other things the jumping of a hurdle. The hurdle is to be one and a half times the dog's height at the withers, or three feet, whichever is less. Thus a Poodle that stands twelve inches at the withers must jump eighteen inches. Most German Shepherds would have to jump the full three feet, since very few of them stand less than twenty-four inches at the withers. A few breeds, such as Basset Hounds, which are built for specialized kinds of work in which the conformation needed for elevation, as dancers put it, is irrelevant, need jump only their height at the withers.

Some dogs can't do this because there is a great deal of careless breeding going on, which leads to the perpetuation of cow hocks, straight stifles and other conformation problems, as well as various arthritic conditions. Furthermore, some handlers can't train their dogs to jump the regulation height on command, even when the dogs are genetically sound. Teaching a dog to work over hurdles is hard. Another problem is that some dogs have been injured while jumping in competition, either because of structural weaknesses, or because in some areas the competitions are held indoors on hard, slick surfaces, or because the dogs are brought into the ring in poor condition. Breeding dogs well is hard. Training dogs well is hard, and running obedience trials competently is hard.

The naive might think that the solution to these problems is obvious. Stop breeding weak animals, and in the case of dogs already alive who are badly built, don't jump them in competition. Find a good trainer to learn from, work hard and in general try to live up to your (human) responsibility to see to it that as many dogs as possible inherit the potential for power and joy that is a dog's God-given birthright. (Roethke wrote,

"This joy outleaps the dog!" marking his awareness of that part of the divine plan that has to do with dogs and leaping.)

But there has been, over the past several years, growing discontent with the jump-height standards set by the AKC, and a move to lower them that is expressed with as much sickly piety as is to be encountered anywhere. The appeal is to justice and democracy. Lowering the standards would confer equal rights on dogs who can't jump and handlers who can't train.

I was standing about the edges of a large meeting in which this view was being hotly elaborated by a woman who was on the verge of tears, so moved was she by the vision of the poor dogs deprived of their rights by the reactionary Kennel Club regulations. With me was Dick Koehler. At our feet was my new Bull Terrier.

"You want equal rights for that pup of yours?" he growled at me. The reply was, of course, "Hell, no!" (Some people never learn.) Behind us, in the training area, a young man worked his magnificent Pointer over jumps five and six feet high. The dog, that is, soared with transcendent ease over jumps twice the height required by the AKC standards. The young man and his dog were overflowing with beauty, and the air around them was full of the light of that exacting joy.

This moment has a background. A year earlier that boy was jerking with sideways bitterness through life, a walking web of cocaine and despair. His discovery, through Koehler's training, of the difficulties of working his dog had given him alternatives.

You should know that there was once a Bloodhound named Nick Carter. He was one of the greatest dogs who ever lived, and his record for cold-trailing and for simply sticking to a trail—105 hours in one case—has not, so far as I know, been equaled. He had to be genetically sound to do this; weak shoulders, poor angulation and so on would have meant pain and inflammation stopping him on the trail long before he got anywhere near the halfway point. Tracking is very strenuous work, physically as well as mentally.

Poor breeding leads to, among other things, a condition

called hip dysplasia, a painful, progressive arthritic condition. Some dogs are so crippled by the age of six months that they can't lift their hind ends off the ground. You have to hold them up so that they can crap. A needleful of sodium pentobarbital is the only alternative most of the time with such animals.

Now I think again of Woolf and her story of Shakespeare's sister. And I am moved to wonder who can measure the heat and violence of a Nick Carter's heart caught and tangled in a dysplastic body.

But what was Woolf telling us? That there is some equivalence between a woman's body and dysplasia? Well, there are some things to notice. For one thing, she doesn't claim that it was the poet Shakespeare's voice that drowned out his sister's. All she tells us is that there was a manager who took pity on her so that soon she found herself with child by him. Now, of course, pregnancy isn't a disease, we all know that, but it does create certain logistical problems, especially if you are sixteen, as the sister was in Woolf's story, and gifted. So consider the manager "taking pity," and consider the woman at the meeting I spoke of taking pity on all of the unathletic dogs.

What I am saying is that one of the things Woolf tells us is that admiring pity doesn't help. Dysplasia is a preventable condition that stops dogs from jumping, and pity is a preventable condition that stops women from leaping.

My friend Eleanor works with disturbed—meaning anything from juvenile delinquent to schizophrenic—adolescents. She became interested in my work with crazy horses and watched with particular interest the effects on young riders when they learn to straighten out their horses.

I became interested in her work, too, and went around with her to the homes of these children, bringing trained dogs with me. I also told the children, during those brief periods when it was possible to speak to them, that I was a horse trainer.

Some of the boys (almost all were boys) responded with interest to this, and Eleanor brought one of them to a horse show, to watch the riding and jumping. He slipped his lead and "stole" a horse, a young stallion who, when the boy opened his stall door, made a wild dash for freedom.

There was a lot of uproar and excitement about this, and Eleanor, aware that in the horse world there can scarcely be a crime more heinous, tried to explain to the boy the gravity of what he had done. I advised against this; the boy was plainly in no frame of mind to comprehend and appeared to be hallucinating. I suggested that in a day or two Eleanor should bring him with her to my stable so I could give him a riding lesson. She did, and I put him on Peppy, who is not only a former holder of the title Craziest Horse in the County but also my most reliable school horse. (I give lessons to a blind friend on Peppy. That's what I mean by reliable.)

When the boy showed up, he was displaying a full range of schizoid behavior. I got his attention and had him mount, and for an hour Peppy took him around, mostly at walk and trot, and carried him over a few low rails. At the end of the hour, the boy was so sore he could scarcely walk—but he was talking, making eye contact, responding to *what was said to him* instead of to the internal voices he heard, or whatever it was that goes on inside. His body was symmetrical, and there was congruence between his movements and his sentences. He was that strange, wonderful thing, a whole, proud, defiant, glorious young human male.

His foster father, one would have said, was the kindest, most understanding man in the world. He never punished the boy for his various difficult behaviors, and was infinitely tolerant and patient. I accompanied Eleanor when she drove the boy home, and when we reached the house he was still glowing, congruent. He started to tell his foster father that today he rode a horse. "And I made him *jump!* A great big strong horse, and I made him jump!"

I was puzzled because the father frowned and somewhat shortly told the boy he had better go take a nap, he must be tired. The boy obediently began to look "tired"—that is to say, spaced out and schizy. But apparently the boy wasn't fully obedient, because I learned the next morning that the foster father had beaten the shit out of him, and he was taken the night before to emergency. This began as an argument between the boy and his father about his continuing to take riding lessons (which I had offered him free).

The foster father said that the beating was necessary and blamed it on Eleanor and me, who confused the boy cruelly just as he was doing so well, taking his naps and so on. That argument didn't stand up in court, as it happens.

There was another young boy named Joey. He had been diagnosed as autistic. At the age of eleven, he had never been known to speak a single word. Joey's therapist had a dog, a German Shepherd, whom she trained under Dick Koehler.

The dog was the only living creature Joey had been known to attempt intimacy with, so the therapist went to Dick, and they discussed getting a dog for Joey and having Joey train the dog with Dick. Dick went out looking for the right dog—it should be a German Shepherd, one that resembled the therapist's dog, and it needed to be emotionally stable and quite bold.

A dog was found and named Jason (by Dick and the therapist). Joey and Jason, with the therapist, started appearing in one of Dick's training classes. Dick treated Joey the way he treated everyone else in class, coming on like an enraged gorilla who has studied the tactics of Marine sergeants in boot camp. He stomped around yelling, trampling on people for blowing it with their dogs.

By the third week of class, Joey spoke his first words. They were to the dog. He gave commands, "Jason, Heel!" "Jason, Sit." Then he advanced to "Good dog!" and "Way to go, Jason!"

Next came sentences like "Jason, you're the best dog of all!" and "Did you hear that? Mr. Koehler says you're super!"

About halfway through the thirteen-week course, he spoke to a human being for the first time in his life: "Mr. Koehler, he chewed my socks. What should I do?"

I doubt seriously that this was a full-fledged case of autism, but it is important to note that Joey's range of autistic behaviors was impressive. In any event, Joey at this point initiated conversation with his therapist. The trouble began at home when he started walking and talking like a human being instead of like a self-destructive lampshade. What he talked about was how much he wanted a dumbbell and a set of jumps, so that he could go on to advanced work with his dog.

I don't know what went on at home in much more detail than that, but the parents started insisting that Dick and Jason were making the child much worse. They pulled the boy out of training class and away from his therapist.

Joey stopped talking. A year later, Dick asked the therapist what became of Joey. The news was that he was hospitalized, drugged and continually restrained because of his attempts to smash his skull open.

I told an uncannily effective psychiatrist about these two children, partly because I wondered if maybe dogs and horses or trainers *were* bad for messed-up kids. He said, "I learned a long time ago, especially when dealing with children who are still in their family situations, not to do too much to threaten the symbiosis if the child can't be adequately protected."

I began this chapter by saying that I wanted the word kindness back. C. S. Lewis has several discussions of the word, including this one, which is in aid of a discussion of the nature of medieval scientific thinking:

> In medieval science the fundamental concept was of certain sympathies, antipathies, and strivings inherent in matter itself. Everything has its right place, its home, the region that suits it, and, if not forcibly restrained, moves thither by a sort of homing instinct:
>
> > Every kindly thing that is
> > Hath a kindly stede ther he
> > May best in it conserved be;
> > Unto which place everything

> Through his kindly enclyning
> Moveth for to come to.*

Thus while every falling body for us illustrates the "law" of gravitation, for them it illustrated the "kindly enclyning" of terrestrial bodies to their "kindly stede" the earth, the center of the Mundus.†

With the aid of this I would like to say that for the trainers I have been talking about, to be kind to a creature may be to be what we call harsh (though not cruel), but it is always to respect the kind of being the creature is, and the deepest kindness is the natural kind, when your being is matched to the creature's, perhaps by a kindly inclining. (Incidentally, when Newton gave us the word "gravity," and in it the word "grave," he gave the law of falling, including the law of falling in love, to death. This made it inevitable, I suppose, that Nietzsche would eventually have Zarathustra declare himself the enemy of gravity. Since in the next chapter I will be talking about Pit Bulls and love, this is a comment on the history of the portion of the texts of both love and Pit Bulls that has become illegible, in part because the law of its love has been given to the grave.)

In this way of thinking, it is *not* kind to let your Pointer roam free to pursue the hunting *Geist*, largely because the handiest game is likely to be the neighbor's livestock, and your dog will be shot. It is closer to being kind, to the dog at least, to make yourself into the kind of person who just feels like training your dog properly and going out after the game with him.

But, of course, there is a problem about whether or not hunting is kind. I suspect not. I suspect that a fight between a Pit Bull and a raccoon is unkind to the raccoon; my evidence for this is that raccoons tend to run away. In the next chapter I will talk about the much deeper perplexity this question raises than any question about whether matching Pit Bulls ought to be outlawed does, although that one is quite deep enough.

*Chaucer, "The House of Fame," II, 730 ff.

†C. S. Lewis, *The Discarded Image* (New York: Cambridge University Press, 1968).

And I am still perplexed, which is why I have told my small stories in place of the philosophy I will someday know more about, a philosophy of the logic of respect and of what happens when respect fails. All of us already know what happens, of course, but what I found myself wanting to have temper tantrums about was a particular form the failure of respect takes. I don't have a name for it, but it tends to be coupled with a lot of talk in which words like "kindness" appear, together with a refusal to imagine the limits of our knowledge of others. And as an animal trainer I cannot help but hear in the stupidities of what is so often called "kindness" the pieties that finance, buttress and sponsor cruelty.

Philosophy often begins with a puzzle. I at one point in my life became profoundly puzzled about the fact that the people I knew—the Koehlers most notably among them—who appeared in oral and published gossip as cruel devils were exactly the people I automatically called on when a child or a dog was hurt, or when I was down and out, or whatever.

I am still puzzled, but the angle and shape of the puzzle has changed slightly, though not its fundamental logic. The puzzle now is: Why is it that I still can't claim that I approve of kindness to dogs and horses?

9

Lo the American (Pit) Bull Terrier

A disproportionately large number of Pit Bulls are able to climb trees.

RICHARD STRATTON

The French philosopher Jacques Derrida, in a lecture on memory and mourning, remarked that we never know, that we die without ever being quite sure, what our proper names are. This is not always so obvious for us, except perhaps in the cases of some newly marrieds. We do not generally feel puzzled or at a loss for an answer when someone asks, "What's your name?" It is, however, obvious in the case of the dog whose name most often these days (in newspapers at least) is the Pit Bull.

There are a number of breeds that are related to the Pit Bull and often confused with it. Among these are:

English Bull Terrier
French Bulldog
English Bulldog
Jack Russell Terrier

Other breeds are accused, in newspaper and word-of-mouth horror stories, of being Pit Bulls, though close inspection of a given incident reveals the individual animals in question to be only very vaguely related to Pit Bulls, if at all. These include Doberman Pinschers, Boxers, Airedales, Rottweilers, certain short-haired European terriers and, in at least one case, a Collie. (In the case of the Collie, who had been written up as a baby-killing Pit Bull, it turned out that no baby had been so much as hurt, though the dog had snapped at the infant. When I protested to the newspaper editor about that, and about the fact that the dog was plainly a Collie, the reply was "But it could have been a Pit Bull!" I thought of telling that editor which breed is actually responsible for the largest number of unwarranted dog bites, but decided against it, and have decided against printing the information here, although most people would find the answer surprising. I decided against it because that would just occasion more horror stories about that breed, and it wouldn't get anyone off the backs of Pit Bull owners.)

The dog I left out of the list above of genuine relatives of the Pit Bull is the American Staffordshire Terrier, which some American Staffordshire fanciers say is actually the same breed, as do some serious Pit Bull people; other members of both groups argue that they have become separate breeds ever since the American Kennel Club opened their registry to the American Staffordshire Terrier. If you own a Pit Bull, or something like a Pit Bull, and are tired of people clutching their purses and babies and shying away from you shrieking whenever they see your dog, just tell them that what you have is an American Staffordshire Terrier. Almost no one, as far as I know, is afraid of American Staffordshire Terriers.

As for the names of the actual dog under discussion. The possibilities include, and have for some time:

Pit Bull
Pit Bull Terrier
American Bull Terrier
American (Pit) Bull Terrier
American Pit Bull Terrier (without the parenthesis)
American Bulldog
Bulldog

If you were to try to write a history of the breed you would also have to work out which, if any, of the following names is a present or past name for the Pit Bull or an ancestor of the Pit Bull. Some of these are now the names of definite breeds, such as the Mastiff and the Dogue de Bordeaux; others are *probably* names for the Pit Bull that have passed out of use. They include:

Irish Pit Terrier	Neapolitan Mastiff
Catch Dogs	Dogue de Bordeaux
Bear Biters	Olde Bulldogge
Boar Biters	Argentinian Dogo
Bull Biters	Tosa-inu
Mastiffs	Colored Bull Terrier
Bull Mastiffs	Bandogs
Molossians	Hog Dogs
Bear Dogs	Southern Hounds

The heraldic and hunting dogs of Western artistic tradition are generally regarded as the best evidence we have of the past of the breed. Indeed, the length of its past is regarded as one of the strongest explanations of its nobility. Thus how a given historian of Pit Bulls or related breeds decides to use any of the above names in defining the breed will be a function of a philosophical decision about what an accurate chronicle is. But you can see, if you compare the lists, that there is a problem. It goes like this in my part of the forest:

As far as the United Kennel Club is concerned, the official name of the breed, which was arrived at after a lot of debating, arguing, divorces, marriages and letters to the editor, is: American (Pit) Bull Terrier. The dogs themselves are called by their affectionate owners one or several of the following: Pit, Pit Bull, Bull Terrier, American Bull Terrier and American Bull-dog. What they actually are, by the way, is: Bulldogs, though that is not the name of the breed but only what the name of the breed is called.

I must digress and say that those breeds of dog that are called Bulldogs (including Handsome Dan, the mascot for the

Yale football team) are not actually bulldogs, or very few of them are. They couldn't get a bull to behave if heaven depended on it for supper. It is the dog who is called the Pit Bull that is the true Bulldog, the one with bloodlines thousands of years old. These dogs, by the way, in a famous painting by Andrea da Firenze, are depicted as guarding the gates to heaven so that only the righteous may pass. The dogs are helping out, in this case, the Dominicans, and the pun—dominicanii, dogs of God—is deliberate.

When James Thurber told his canny and charming tales about Pit Bulls he called them Bull Terriers ("American Bull Terriers," he says at one point, "none of your English Bulls"). In Thurber's day there were no horror stories of any significance about Bull Terriers eating up old ladies and nubile maidens. You don't have to be much of an historian to know this, you have only to reflect on what sort of writer Thurber was. I can't imagine him passing up the opportunity to poke fun at the tellers of those stories, but in his day people amused themselves by telling the stories about Bloodhounds. This may seem queer to us today, because of course everyone "knows" now that the Bloodhound is a sweetheart and a patsy. The reason everyone "knows" is because of Thurber—his drawings, his tales and his piece "Lo The Gentle Bloodhound!" This is one of the rare places, by the way, at least in the Thurber that has been collected between book covers, where anger threatens to destroy his dangerously mild-mannered wit.

I have, as it happens, had what we in the twentieth century are so fond (rightly, I think) of claiming as an essential element in the search for truth—actual experience of owning a Bull Terrier and thus of coping with the horror stories and the maddening ignorance that backs them, an ignorance that is insisted on as though it were a virtue. This is from time to time backed up by threats of various sorts, including threats against the life of my dog and in one case against the job security of a friend who allowed me and my dog to visit her in her office. The threats come from tireless good citizens who manage to hold, simultaneously, such thoughts as:

People who are kind to dogs are kindly.
People who are kind to Pit Bulls are evil and bloodthirsty.
Dogs are good, loyal and affectionate.
Pit Bulls are dogs.
Pit Bulls are evil.
Killing dogs is evil.
It is good to kill Pit Bulls.

But I want to avoid, if I can, fulminations against what Auden called "the complicated, accurate and false details," though some of them come into it. I will try instead to make as true a fiction as I can.

In, as I recall, the early seventies, the first of the current versions of the horror stories about Pit Bulls appeared. I didn't see the original one, which I am told was the product of the inflamed mind of a journalist in Chicago, but as the story was picked up and reprinted, revised and improved by nearly every paper in the country as far as I could tell (I was doing some traveling then), I got to read it often. The stories had headlines with words like "maul" and "massacre" in them, and connected the killing of infants, little old ladies and young girls with the fact that evil and bloodthirsty men cruelly trained these dogs to fight, starting them off with small baby animals put into sacks that hung above the dogs' heads. Movements began throughout the country to have the dogs killed, licensed, outlawed.*

At first I was only mildly amused and not especially worried, because I have my real life in a world in which there are no

*I ought to say that one of the recent horror stories that appears to be true is one in which a Pit Bull kept chewing on the bumpers of a neighbor's car. The car owners didn't mind this particularly, as it wasn't much to look at as a car anyway. But then they got a new car, a fancy one. This one—I forget the make— had rubber bumpers, so of course the dog removed them entirely. One news account of this said that the car owners were not mad at the dog but rather at the car company, for charging so much for a car that wasn't dogproof. Then there is the sadder one about the woman in Chicago who was bitten by a real Pit Bull, but she knew enough about dogs to understand that it was stupid handling that had caused the bite, and she wanted to adopt the dog. But the dog was, alas, "executed," as the headlines had it.

horror stories about Bull Terriers. Indeed, they are generally recognized as such an amiable, easygoing lot that Dick Koehler has wondered if they wouldn't be too "soft" in particular ways to work with husky autistic children. Among fighting-dog men, the ones who know what they are about, a man hater is almost never bred and isn't likely to survive anyway, because man hating tends to go with cowardice in the pit. And a United Kennel Club Pit Bull show is probably the safest kind of dog show to attend, since aggression toward either people or other dogs not only disqualifies an exhibitor from winning, it leads to expulsion from the show grounds. Hence there are very few biters among Pit Bulls.

Of course, they are very strong, so if they do bite you, you sit up and take notice, but that is true of any breed worth the name "dog." I have seen hundreds of these dogs and not one bite except in the course of duty, or as the result of the phony "attack-dog training" that messes up Pomeranians, St. Bernards and everything in between. I myself have been bitten by a Schnauzer, several Cocker Spaniels, a Poodle, a Labrador, a Samoyed, a St. Bernard and various other breeds, but never by a Pit Bull. I might have been, I suppose, but as it happens I wasn't. If these dogs have a flaw in their relationship to people, it is that a few of them—not many—have a tendency toward reserve, or a kind of aloofness that is a consequence of their often being the kind of dog who loves above all reflection and meditation, and so they tend to hang back in social situations they don't understand.

Not in fighting situations, of course. You have to understand about fighting dogs or hunting dogs who take on opponents like mountain lions—any dog in which the quality called "gameness" matters. In a true fighting dog there is no ill temper, nothing personal held against an opponent, no petty resentments of the sort found in the snappy, ignoble animals so beloved of people who want to kill Pit Bulls. I had, for example, an Airedale who was a visionary fighter, a veritable incarnation of the holy Law of the Jaw (Never Let Go). The way you could tell that Gunner was going into fight mode was by a certain precisely friendly wagging of the tail, a happy pricking of the ears and a cheerful sparkle in the eye that quickly progressed to an expression of high trance. He was, when he wasn't fighting or thinking about fighting (he didn't think about it all of the time, only when it was appropriate), a dog of enormous charm and wit who never minded playing the fool.

One of the things he liked to do was to climb up the ladders of playground slides and then slide down, with a goofy, droll look in his eyes, his ears flying out like the arms of a child playing at being an airplane. His charm was often an annoyance; he always insisted on Making an Entrance and looking around happily for the cheering section.

The only time I knew him to menace a human being was when he was about a year old and late at night a man attacked me with a knife. That man lost part of his nose and cheek and I don't know what else (it was dark). Gunner lived for over ten years after that and never again so much as growled at anyone, except a rather aggressive evangelist preacher who grabbed at me one day when I was walking to the library on campus.

This brings me to an important aspect of the structure of the horror stories. After Gunner's attack, one response I got from a few (middle-aged, male) friends was that I had no business harboring a vicious dog like that. I turned the word "harboring" over in my mind but could make no sense of it and countered by weakly offering the view that most of the time people had no business pulling knives on me, but of course that didn't wash. I was informed kindly, gently and firmly that if women would just stay at home at night these things wouldn't happen. Well, we all know all about that these days, but the interesting thing is that these conversations took place in Gunner's presence, because I took him with me virtually everywhere—like a purse, a friend of mine complained. But, curiously, it was never Gunner who was supposed to be the author of the bloody deed, even though I said quite clearly that it was. It was vaguely supposed to be some other Airedale, one I kept locked away in the cellar for my own dark purposes and systematically trained in ever-increasing spirals of viciousness. Horror stories are about things that have neither presence nor being; I doubt that one could be told effectively in the presence of its putative "subject." Husband-wife quarrels may appear to be an exception to this, but in the case of couples, of course, the storyteller never gets a chance to get to the point.

As it happens, I have never attack-trained a dog and never been involved, even as a spectator, in an organized dog fight. But the fact that I own a Pit Bull makes me part of the horror stories, and the other thing about the subject of a horror story is that the subject is the one who knows the *least* about himself or herself of anyone in the world. I tried telling someone a few months back that I had never attack-trained a dog or seen a pit match. The someone replied knowingly, seeing through to the depths of my bloodthirsty soul and forgiving me for it, "But you'd *like* to, wouldn't you?"

I used the word "visionary" a few paragraphs back to talk about my dog who was a genuinely game fighter. To understand more of what I mean by that, consider James Thurber's account of a visionary dog fight in "A Snapshot of Rex." Rex is described as a dignified and powerful dog who seems to have been inspired by virtually any task if that task had sufficient—well—*taskness* to enable Rex to use his tremendous power. Retrieving ten-foot rails with his mighty jaws, swimming against strong currents for the sheer fun of it, bringing home from a distant spot a chest of drawers is the kind of thing Thurber tells about. So for Rex fighting was no doubt just another chance to use himself fully, to bring his full creatureliness alive. He also had another kind of power, a kind that a lot of dogs have, but Bull Terriers have it sometimes to quite an extraordinary degree. That is the power to compel devotion on the part of humans. Thurber says of his family that when Rex's fighting caused relatives to look "upon [it] as a blot on the family name," and tried to persuade them to get rid of Rex, it was no go. "Nobody could have made us give him up. We would have left town first, along any road there was to go." Rex

> never started fights. I don't believe he liked to get into them, despite the fact that he came from a long line of fighters. He never went for another dog's throat but for one of his ears

(that teaches a dog a lesson) and he would get his grip, close
his eyes, and hold on. He would hold on for hours. His
longest fight lasted from dusk until almost pitch-dark, one
Sunday. It was fought in East Main Street in Columbus with
a large, snarly nondescript that belonged to a big colored
man. When Rex finally got his ear grip, the brief whirlwind
of snarling turned to screeching. It was frightening to listen
to and to watch. The Negro boldly picked the dogs up
somehow and began swinging them around his head and
finally let them fly like a hammer in a hammer throw, but
although they landed ten feet away with a great plump, Rex
still held on.

The two dogs eventually worked their way to the middle
of the car tracks, and after a while two or three streetcars
were held up by the fight. A motorman tried to pry Rex's
jaws open with a switch rod; somebody lighted a fire and
made a torch of a stick and held that to Rex's tail, but he
paid no attention. In the end, all the residents and shopkeepers
in the neighbourhood were on hand, shouting this, suggesting
that. Rex's joy of battle, when battle was joined, was almost
tranquil. He had a kind of pleasant expression during fights,
not a vicious one, his eyes closed in what would have seemed
to be sleep had it not been for the turmoil of the struggle.

This event, as the narration gives it to us, is an allegory of
the usual relationship, in my world, to dog fights. No one
wants them especially, and when they occur, they are framed
by attempts to stop them—quite imaginative and strenuous
attempts—but at the center there is nonetheless awe and admi-
ration in the presence of a beautiful and nearly pure cynosure:
when Bull Terriers fight, what we see approaches a Platonic
form. We are compelled by dogs and dog fighting (whether or
not we hate them) in roughly the way the tail of a dog is
compelled by the dog to wag, which is no doubt why our very
word for a center of attraction or interest—"cynosure"—has its
roots in the Greek for "the tail of the dog."

The story of Rex is a portion of a long, ancient and noble
text of the Bulldog, but that text has lately become illegible,
so illegible that we no longer seem to know what sort of text
it is. We are misled by a number of things, not all of which I

can go into, but one of them is a certain version of Romanticism. The Romantic hero is a dangerous hero, one who of necessity stands outside of the social order for the sake of heroism. Heroism in this view bows to nothing, including (if the hero is, say, an artist) its own inventions. Regarded from within the social order, the Romantic hero is by definition a monster.

But Rex and other good Bull Terriers are not Romantic heroes. They are much more like Spenserian heroes, operating in a situation in which, as in the Fifth Book of *The Faerie Queene*, Justice has literally fled the world in disgust, with the consequence that Artegal, the Knight of Justice, is continually confronted with the problematics, in an incoherent world, of knocking heads together. Hence he sometimes cannot take heroic action, as for example against the Giant of False Justice. However, Talus, Astraea's (Justice's) groom, whom she sent to be Artegal's squire, can and often does—and, incidentally, is often described in doglike terms. Talus must sometimes be restrained, as must dogs. So the violence of a fighting dog, or a police dog, or for that matter your own dog guarding the proverbial baby on the blanket, stands in relationship to our own violence in a domestic (non-military) situation roughly in the same way that Talus stands in relationship to Artegal and to all of humanity. I mean that a police dog or a Bull Terrier taking down an intruder can act with a degree of moral clarity that eludes us most of the time. I think that certain reports I hear of triumphs in university committee meeting fights (with appropriate horror stories about the opponents, of course), like similar tales I used to hear in a bar I liked which was frequented by cowboys and truck drivers, are expressions of a deep yearning we have for that sort of clarity, which is why the arguments for killing all Pit Bulls sound so much like arguments for more and better dog fights.

I have not, of course, answered the question about whether Bull Terriers really do eat people up when they ought not to at a rate that is statistically significant. I certainly don't have a hawk's-eye view of the world and the dog stories in it, whether they are Thurber's stories or those that have their *locus americanus* in the writing of, say, a Chicago journalist. In fact, what I have is a kind of view from under the belly, so that is the story I have to tell.

In order to tell my story, I have to return to the stories I started hearing in the early seventies. One of the favorite elements in these stories is a gleeful account of how Pit Bull puppies are trained to be killers by starting them off on declawed kittens. The interesting thing here is that an authentic and intelligent admirer of good Pit Bulls like Richard Stratton* finds this to be an insult to the breed and the Pit men, partly because most lovers of Bull Terriers turn out to be saps about animals of all sorts (they hate hunting quite often) and wouldn't dream of such behavior, and partly because they have a kind of Nietzschean sense of what counts as a worthy opponent that makes kittens, declawed or not, exempt. Someone like Stratton has deep contempt for anyone who would set a Pit Bull against a dog that was not a match, even if the dog was a 120-pound specimen of the Japanese Tosa, bred specifically for fighting. That's a kind of braggadocio that has no place in this world.

Beyond that, one source of Stratton's anger against the declawed-kitten talk should be noted. What he says is "We're talking about a dog that can stay the round with a porcupine. He doesn't need to practice on kittens!" That is, the charge of cruelty to kittens is a subportion of the more serious charge, the insult to the nobility and courage of the breed.

I should interrupt myself again and add that the horror stories are beginning to come true. Ignoble examples of *Homo sapiens*, entranced by the stories, are going out and getting themselves dogs that may or may not be genuine Pit Bulls and giving them declawed kittens in sacks and God knows what else, and encouraging them to kill them. Similar types are finding the few—very few—man-biting Pit Bulls and deliberately breeding them. So, for different reasons, the same thing is happening to the Pit Bull that happened to Collies, Cocker Spaniels and German Shepherds; their popularity is producing

---

*Author of, among many other books, *The World of the American Pit Bull Terrier* (Neptune, N.J.: TFH Publications, 1983), staunch battler in court and highly respected show judge.

some dogs that are real curs. But the Cockers are recovering, there are some stable-hearted Shepherds being bred without inborn arthritis, and Bull Terriers will recover, too, as they seem to have for thousands of years in one form and another.

Richard Stratton in a number of places writes at length about the development of the horror stories and their consequences, which include cases of good citizens seeing to it that other people's dogs were impounded and destroyed. One of the most appalling of such cases, in San Diego, involved the killing of an entire line of dogs on whose development the breeder and handler had spent decades. Later, a court ruled that the killing of the dogs was unconstitutional, but the corpses of the dogs appear not to have been impressed by this development. Having given a detailed description of this increasingly popular form of humania, Stratton writes:

> Later there followed the proliferation of foolish felonies in several states. In each case the approach was the same: the same stories as before were told, to which was added that certain states have very effective laws.... Each state was assured that it was the center of dog fighting in America, and wasn't that a shameful "honor"? A news-media blitz characteristically preceded attempts at putting through legislation. In some states, penalties as high as ten years in prison were specified.

Well, that, in outline, is the situation. This morning (I am at the moment in Greenville, Illinois) I heard someone insult someone else by accusing him of owning a Pit Bull, and the demeanor and attire of the opponents made me suspect that the fact that dedicated Bull Terrier men are often Southern, rural, and wear clothes that just wouldn't fit in in Manhattan has something to do with some of the righteousness behind the legislative attempts. The Civil War is still on, though the location of the Mason-Dixon line isn't as clear now as it used to be.

Before I got my Pit Bull, I knew, of course, about the scare stories, and some news of the legislative successes had reached me, but I don't actually talk much with people who tell such stories, and I thought vaguely that it was just a mildly annoying new way the humaniacs had found to amuse themselves. I happened to be looking for a good working dog and put the word out among the dog people I knew. There was a wide variety of breeds that were acceptable to me. Since I am not fond of grooming, some good dogs like Standard Poodles, Bouviers and the like were pretty low on my list but not out of the question. Dobermans and Boxers were pretty high, as were all of the fighting breeds. Airedales need a lot of grooming, but they are always high on my list, and so on. I was just waiting for a dog with genuine class to show up. I would have looked at a Cocker Spaniel if a reliable informant had told me of a good one.

I heard, eventually, of a litter of puppies in which there was a promising little bitch. They were Pit Bulls and within my price range, so I went to take a look. The bitch puppy I had been told about looked as good in the flesh as she had in rumor. I bought her and named her Belle, a name that may sound fancy to Yankee ears, but a good old down-home name for a nice bitch. In Belle's eyes there is a certain quiet gleam of mischief and joy but more strongly a general air that makes it clear that with this pup you are going to be dealing with her on *her* terms, and they may include the odd impulse to make a fool of you, though she wouldn't offer to play Bulldog-style with someone she thought didn't understand the forms of Bulldog play. In this, she has more judgment than my Airedale had.

Belle is a red brindle, with some white patches here and there, which sometimes give her a raffish air but at other times, when she has her dignity about her (about 99 percent of the time), make her look like the queen of an exotic and powerful nation. She is fairly typical of her breed in that she is very serious about whatever she happens to be doing or being. Especially sleeping. I've had her going on three years now, and the most violent thing she has done to date was, one day when her pillows were in the wash, to go about the house scouting and appropriating everyone else's pillows. Not *all* of the pillows, only the newer, plumper and more expensive ones. (She was quite young when she did this, I hasten to add—maturity has brought with it a sense of the importance of respecting other people's property rights.)

As Belle got a little older, she began taking an interest in the welfare and development of James, my year-old nephew, whose first word, incidentally, was "doggie!" When James threw a plaything out of reach, Belle would bring it back to him. James was entranced by this and spent most of his time throwing his playthings out of reach. Belle continued patiently, with a worried look about her, to fetch them for him.

I must remind you of the seriousness of mind of this breed. It became clear after a while that Belle was not just "playing fetch." Bull Terriers are never *just* doing anything. She began bringing the baby her dumbbell—*not* a plaything in Belle's code—and attempting to get him to handle it correctly, fussing about his future development. This is only natural; Belle's mother was infinitely devoted to the education of Belle and her littermates, so Belle takes her responsibilities seriously. She seems to feel a necessary condition of fully developed human-hood is dog-training skills, and it suddenly dawned on me that she was trying to teach James to train *her* to retrieve! In fact, her dedication to this project was somewhat annoying, as though she were a nanny who never regarded herself as off duty and let you go and play with your own kid.

Belle's behavior with James is related to a standard Bull Terrier trait. If purity of heart is, as Kierkegaard said it was, to will one thing, then Bull Terriers have purity of heart. A less generous way of putting this is to say that they have one-track minds. Bill Koehler quite seriously warns Bull Terrier owners not to play ball with their dogs in the house, except on the ground floor, because if the ball goes out of the window, so does the Bull Terrier. Indeed, I was taking a walk one day with Belle, which is a very agreeable activity because of her meditative, philosophical bent. Suddenly, she simply disappeared. I was quite worried, as she is not the sort of dog to go off for no reason, forgetting her responsibilities. I found her fairly quickly, at the bottom of a fifty-foot drop near the path we were on. When I peered over, she was on her back, but she quickly rolled over, got up and shook herself with an exasperated sneeze and trotted around and up, back to the path and me. Clearly, something had gotten her attention, and fifty-foot drops are not the sort of thing that interfere when something has a Bull Terrier's attention.

I was talking with Dick Koehler one day, saying how nice it was to have Belle around, but how hard it was to explain why. Dick said, "Yeah, it's hard to explain. They are so *aware*." And that's it, that's the quality that Belle radiates quietly but unmistakably; awareness of all the shifting gestalts of the spiritual and emotional life around her. She spends a lot of her time just sitting and contemplating people and situations (which is one reason some people are afraid of her). Since in her case this is coupled with a deep gentleness—no bull-in-the-china-shop routines, once puppyhood was past—Dick urges me not to have her spayed, for a while at least.

The reason is, he thinks she might be a good foundation dam for a line of dogs bred specifically to work with the handicapped. This brings up yet another aspect of the dog horror stories. They tend to be about just those breeds, including, for example, Dobermans, who are the best prospects for work with, say, schizophrenic or autistic people, the old, those in wheelchairs and so on. Some readers may remember the horror stories about German Shepherds "turning on their masters"—dogs with whom the safety of the blind can be trusted! I think that the same qualities that make these breeds reliable companions for the more-difficult-to-care-for-and-live-with members of our species inspire the horror stories. In Belle's case, her refusal to play with strangers who come cooing up at her, which sometimes causes the strangers to fear her, is the quality that would make her reliable in a distracting situation if her quadriplegic master really needed her attentiveness.

Most dogs have an unusual amount of emotional courage in relationship to humans, being willing and able to keep coming back, having the heart to turn our emotional static back to us as clarity when they can. But dogs who work with people with various disabilities, including the sort not always listed under pathology, need much more of this quality in order to do a proper job of being a dog. At best, someone who is, or who perceives him or herself as, powerless will be querulous from time to time and may be downright loony in his handling of the dog. The dog who can keep her cool and continue to do her job properly under such circumstances has to be something more than just cuddly and agreeable, and certainly mustn't have any heart-tugging spookiness in her makeup; such a dog must be prepared to *think* and act in the absence of proper guidance from the master and (as in the case of Guide Dogs) in the face of the wrong guidance. Hence, such a dog must be able to give the moral law to herself when her master (who, of course, runs the universe from the dog's point of view) fails to act on the law of being.

An interesting consequence of this is that the qualities that make a dog a good prospect for work with the handicapped are like the traits that make a good police-dog prospect. (I am not talking about some of the phony "training" of either police dogs or dogs for work with the handicapped that goes on in some parts of the country, by the way.)

The following story is a clear case of a dog giving himself the moral law. It's a true story, in the historian's rather than the poet's sense of "true," but I have changed the details so as to make the characters unrecognizable. Fritz, a Doberman, was the partner of Philip Beem, who can't be described as anything but a bad-assed cop. (There are plenty of policemen to whom this description does not apply, but Phil was not one of them.) I knew both of them somewhat, having worked with them in training situations, on tracking and scouting. When he was working his dog, Phil was pretty human and in general a better police officer. This is a fairly frequent phenomenon when a policeman is working a good dog and a properly trained one, and I suspect it is connected with what I was saying earlier about Artegal, the Knight of Justice, and the reasons he needed his squire Talus to do some of the head knocking. Policemen have their being on the frontiers of the social order, and that is a very uneasy place to inhabit. Having a good dog for a partner, a dog whose judgment can be relied on, relieves some of the burden. Besides, it's dangerous doing things like scouting warehouses for desperadoes, and dogs are a lot better at finding them in time than we are. A scared cop, like a fear-biting dog, is dangerous. But this is not a story about that.

One night, Officer Beem stopped a young black woman for jaywalking and started clubbing her with his nightstick, for the sheer fun of it as near as anyone could make out. (There were witnesses.) Fritz attacked—not the woman, but his policeman partner, and took his club away from him emphatically.

Now Fritz was not only by nature a good dog, he was well trained and had a keenly developed sense of what his job entailed, what did and did not belong in this particular little dog-human culture. Sitting by while people got beat up for no good reason was not part of his job, it simply didn't belong. While it would not be exactly wrong to interpret this story by saying that Fritz was moved by compassion or a sense of rescue or protectiveness, it wouldn't be quite right, either. He simply knew his job, had his own command of the law in a wide sense of "law" and was putting his world back in order.

Initially, Phil claimed angrily that the woman was resisting arrest and the dog had "turned on him." "Never did trust those damn Dobies anyway." The police chief, fortunately, knew something about dogs and people both, and Phil is not only no longer talking that way, he is working well with Fritz and is on his way to promotion. That part, however, didn't get into the papers, so the Dobie horror stories were given a boost. It was, of course, Fritz's capacity to love Phil that made this into a story with a happy ending, but for such a dog, love doesn't make a whole lot of sense outside the context of a discipline, a discipline in the older, fuller sense of that word in which the context is the cosmos and not a classroom. Love has teeth, I keep trying to say.

Pit Bulls are this way often, only even more so. When Belle was not yet five months old, one afternoon while I was abstractedly working at something, I was startled into consciousness by her suddenly giving out, in place of the wimpy, puppy bark I had so far heard, a full-fledged, grown-up, "I've got duties around here" guard-dog bark.

Investigation showed that the gardener had arrived and was going into the backyard by the side gate *without asking permission*.

So I said, "What's up, Pup?" and put her on her leash and followed her outside to check the situation out. (This is part of the handling of a natural protection dog, a procedure designed at once to encourage the dog's protectiveness and respect it, while making it clear that she must think and exercise judgment.) When we got outside, I said, "Oh. That's just the gardener, and you don't have to worry about him," and, putting Belle on a stand-for-examination, I asked the gardener to pet her.

He refused, saying that he was afraid of her. Even though he was a shifty-eyed sort that has good reason to be afraid of dogs, this worried me a bit since Belle was only a puppy, and while it wasn't too early in her career for her to be barking at strangers who enter the premises in an abnormal manner, she was too young to be seriously menacing anyone. So I asked if she had ever done that, offered to bite him, or whatever.

He said, no, she'd never bothered him at all, but the thing was that he carried liver treats with him on his rounds, in order to "make friends" with the dogs. The only dog—including a couple that had had some form of formal attack training—who had refused his liver treats so far had been Belle. No, ma'am, she didn't growl or anything, just turned her head away.

I refrained from telling him how rapidly anyone who would offer a bribe to a Bull Terrier sinks in the dog's estimation, really plummets to contempt and suspicion, and simply suggested that in the future he knock on the front door when he came to work, and I would make sure the dog was in the house. Since then, Belle, understanding the situation, announces his arrival with two precise barks and otherwise seems content to let him do his job—though she does keep an eye on him.

I was, of course, full of dog-owner pride but also uneasily aware of the responsibility I had taken on in this dog who needed no training to know a bribe when she saw one, a dog who could give the law to herself. I don't mean that I have the slightest fear that she would ever bite me—that isn't part of her law or her metaphysics—but any unfairness or sloppiness in my handling would, I knew, be made known to me. Belle has an aristocratic genius for expressing emphatic approval.

These qualities, which Belle has and which Fritz had, that make them able to act with moral clarity are all part of genuine love. And although they may not know what they are talking about, people who say that the handicapped, the aged, the delinquent and the mentally retarded need love are right. But they need the kind with teeth. So now it looks as though we tell horror stories about dogs and spouses because it is love that horrifies us. (Consider the film *Cujo*, in which a wife's infidelity seems to lead directly to the horror and the terrifying battle against it. The horror in this case is a rabid dog, a St. Bernard, who before he went mad was wonderful with children and so on.)

Another side note: Belle's behavior with the gardener demonstrates her restraint and her good judgment, but the incident will feed the horror stories in that town.

In the meantime, of course, I have been training Belle and enjoying things like her work with the baby, coming to know in detail the implications of her deep awareness and understanding. And, in training her, I am astonished at how easily it goes. This is not unusual with these dogs; a lot of my friends speak of having the sensation that they aren't so much training the dogs as reminding them of something. However, in other dog circles, people debate about whether it is so much as possible to train Bull Terriers at all. This is because they are unresponsive to anything short of genuine training. Belle is as honest as daylight in her work, which means that my training technique improves a lot, since she simply doesn't respond if it is done wrong. She is committed, and she expects me to be. Thus, it is easy to mess up these dogs precisely because they know so much about how it ought to go. Once, I picked up Belle's leash and some other equipment, preparing to take her outside and work her. Then a conversation, a trivial one, distracted me, and she barked three times, sharply, to remind me of my duties. She doesn't do this when my attention is on something important.

One of the things I taught her was that before she goes through any door to the outside, she has to sit and wait for the release command. This was easy to do, as Belle takes to the forms of domestic order. Then, when she was still young, I was out of town for a week, leaving Belle behind in the care of a friend. My friend is a splendid woman, no two ways about that, but she has never seen the point of all of the time I spend educating, as she puts it, the poor dogs, who would rather be left alone.

When I got back from my trip, I heard the report that Belle, no matter how hungry or thirsty she was, and no matter how full her bladder was, wouldn't go through the door for my friend. My friend would simply swing the door open and expect Belle to skip through, not honoring Belle's sitting as a signal for domestic discipline. When my friend coaxed and cooed at her, Belle would lie down flat, looking depressed, ears and tail low and immobile. (This melancholy imitation of the Rock of Gibraltar is Belle's usual response to coaxing, flattery and insults, and it was to cause me some social problems later, as we shall see.)

I responded to this by canceling another upcoming trip, figuring that later, when Belle had some more experience under her belt, she wouldn't be so vulnerable to the various incoherencies that came her way, since maturity tends to bring confidence with it. At this stage, it was possible to break her heart, and a brokenhearted Pit Bull was not something I wanted to have around. My canceling the trip, by the way, is less a comment on what kind of temperament I have than on Belle's, and on that power I mentioned earlier, in connection with Thurber's tale of Rex, of the Bull Terrier's compelling devotion on the part of humans. That is why the ladies and gentlemen who want to exterminate Pit Bulls, although they may win some battles, will never win the war. They haven't the right sort of dogs to help them think the situation through.

It is now time for me to say emphatically that my praise of this breed should not be construed as advice to rush out and get a Pit. They do like to fight other dogs, and they are in many ways a tremendous spiritual responsibility. If you're ready for it and can find a *real* dog trainer to help you figure out what you've gotten hold of, then go to it. I suspect that if Nietzsche had known Dick or Bill Koehler and had had the right Bull Terrier to train, he wouldn't have had to go mad and might even have learned to have a love life. His dog, of course, would not have allowed his sister's tyrannies.

But be prepared. If your boss comes over for dinner and talks baby talk to your dog or perhaps offers her an hors d'oeuvre, and the dog regards him impassively or turns away, the boss's feelings will be hurt, and your job may be in jeopardy. Also, if the boss later gets tipsy and tries to insult your dog, he will get the same treatment. And be sure your spouse or lover is not the sort of person who will have his or her feelings so hurt. The dog, remember, has the power to compel your loyalty.

Beyond this, if you should succeed in, say, training your dog to track, you're in another sort of trouble. You may think that after four hours in the high desert in August, tracking the lost child, it's time to take a canteen break, but your Pit may think otherwise, and Pits have a tendency to win going away at weight-pulling contests. Also, if the child went through the cholla patch, so will your dog, and so will you.

Any laziness in training may cause your Bull Terrier to decide you no longer exist, and some of these dogs can make things happen just by concentrating hard enough. Even if that doesn't happen, it is more than a little unnerving to have a Bulldog walking around the house or napping as though both the house and the neighborhood were uninhabited.

In my case, aside from the digression into loss of faith when I took my out-of-town trip, things with Belle are becoming more and more serene. Or were until I began taking her to the university with me. She was still a puppy, and not a very big one, the first day I went into the department office with Belle at heel. One of the secretaries was so struck with terror that she couldn't speak, and I couldn't figure out what was wrong because I had, as usual, forgotten about the horror stories. A friend came in, assessed the situation and asked, "What's wrong, Frieda?"

"Tha . . . th . . . that . . . *dog!*"

My friend looked Belle over for a moment and said, puzzled, "But that's only a puppy!"

"That doesn't matter with those dogs. They're born killers!"

Belle was by now regarding the secretary in uneasy puzzle-ment, since she didn't know anything about the horror stories. But now she has had her first lesson. (I suspect that some Bulldogs, once they work it out about the horror stories, do start biting people who are sending out the wrong brain waves. So would you if things went as they went for Belle. She, as it happens, didn't start biting, and very few Bulldogs do, but I wouldn't have blamed her if she had.)

For months, whenever Frieda's path and mine crossed on campus, Frieda sidled along the wall, as far from Belle as she could get, or ducked into the nearest open doorway until we were safely past. She behaved, in short, like a very guilty woman, and dogs, like people, figure that evasive behavior of this sort is suspicious. So Belle is now more wary than she would otherwise have been. In fact, people who are irrationally afraid of dogs behave in almost exactly the same way as profes-sional "agitators"—people who, in the course of attack work, teach dogs to be suspicious. I happen to know how to teach Belle to read and respond to this situation correctly and peace-fully, and she has a lot of natural judgment anyway and learns to distinguish between idiots and real hoods, but things can go otherwise.

Other weird things happened. A campus policeman, a mem-ber of the K–9 Corps who had a police dog with him, started following me about from time to time but never actually made contact. Eventually there was a phone call from the police chief. The officer was worried that my Pit Bull (who has never so much as looked twice in his dog's direction, having other things, like my colleagues, to worry about) would kill and eat his police dog. I, instead of explaining that dogs can be trained (since those K–9 dogs weren't what I call trained, they wouldn't know what I meant), got hysterical about the rising rape rate on campus, and a few other things I knew about that were an embarrassment to the police department, and from then on they, at least, left me alone. I would rather have been able to talk with them about police dogs.

Another friend who, though he trusts me enough to semi-relax in Belle's company, said one day, "I always get the feeling that dog is sizing me up as a bite prospect."

Well, Belle *is* sizing people up. Not as bite prospects, but as problems in moral philosophy and metaphysics. They tend, after all, to behave in a rather bizarre fashion, and she can't rest until a characterizational problem has been solved. So when someone descends upon her squealing, "How cute!" and stoops to fondle her, Belle draws back a little, not haughtily but rather worriedly, as though she senses the dangers of getting entangled in that rhetoric. The nice person's feelings are hurt. In another case, when a more sensible person stooped to fondle Belle and got the same response, she just stood up and said, "Oh, I'm sorry, I should have realized I don't know her well enough to get that familiar." It is men, by the by, who are more apt to be permanently offended by Belle's aloofness, but this is not their fault.

As I have said, the horror stories extend to me. Belle is plainly the outward and visible sign of an inner viciousness. Some of the expressions of this get back to me. "Oh, yes. Vicki Hearne. She has a very repressive ideology. She keeps Pit Bulls, you know." (I didn't know I had any sort of ideology and am entranced by this.) Also "Vicki is a threat to the collegiate atmosphere with that dog of hers." This is probably true, since I don't in any event know what a collegiate atmosphere is.

Of course there is "She *delights* in harboring vicious animals." There's that word "harboring" again. I thought I was rid of it. On one occasion a man stopped me in the hall—a man I knew only slightly—grabbed my arm and said, "Young woman [young woman?], there is something to those stories about Pit Bulls, and you have no business bringing such an animal here." I wondered about that one a lot. I mean, if the dogs are so dangerous, why did he feel free to take liberties with my person when Belle was at my side? In fact, Belle or no Belle, whence those liberties? I didn't reply because I couldn't think of anything to say and walked on, followed by "Won't listen to reason!"

The father of an autistic child said in a book about his boy that if you have a crazy kid you find out what the actual web of society is. Things are not nearly so bad for the owners of Pit Bulls as they are for the parents of autistic kids, but there are some feeble parallels. There's the eagerness to get the kid or the dog out of sight somehow or other, and there's the freedom so many people feel, and that parents of messed-up kids describe so wearily, to tell you what your own life is and what to do about it.

In time, though, Belle herself effects changes in the stories, because the serenity and sweetness she radiates are so strong that they are felt by all but the most distant of the tale-bearers. (Are they tales? This is part of the question about whether a rumor is a text.) Now what I start hearing is "Vicki, I don't know where you get off thinking that's a vicious dog. That dog wouldn't hurt a butterfly, a real patsy if I ever saw one!" Or "Vicki likes to think she's tough, but I'll bet she can't bring herself to give a grade lower than B+, and just look at that mushy dog of hers!"

It is this, the way the horror stories flip over, that suggest that we are on to something. "That dog wouldn't hurt a butterfly" is a part of the same logical structure as the "born killer" part, an insight I owe largely to Cavell, who writes:

> The role of Outsider might be played, say in a horror movie, by a dog [who] allegorizes the escape from human nature...
> in such a way that we see the requirement is not necessarily for greater (super-human) intelligence. The dog sniffs something, something is in the air. And it is important [to the horror] that we do not regard the dog as honest; merely as without decision in the matter. He is obeying his nature as he always does, must.

It is important to tellers of dog horror stories that "we do not regard the dog as honest; merely as without decision in the matter." That is the hidden and stinging part of the logic that requires these stories. An Outsider is required, somewhere or other, and the creation of the Outsider is one of the functions the so-called philosophy of animal consciousness is usually accomplishing with its bizarre proposals of what Richard Jef-

fries, in a letter to me, called the "bogus" issue of anthropomorphism. That seems right to me, too, but if it is a bogus
issue, then what would a real one be? Consider the falseness of
"wouldn't hurt a butterfly" in Belle's case. As it happens, Belle
would nail anyone who threatened me seriously. Notice that I
said *seriously*—she didn't do anything to the guy who grabbed
my arm. She does have judgment; I do regard her as honest
and not as "merely without decision in the matter." She is not
obeying her nature in the way, say, that a falling stone is
obeying its nature as a bit of matter. She is not morally inert.
But neither am I. Would it be right to say of her or of me that
in perceiving, believing or deciding anything at all we are
disobeying something? Our natures? And where does this leave
me in trying to decide what the differences are between me and
Belle? Trainers devise tales and philosophies based roughly on
versions of *Paradise Lost* and say, in effect, that it is the human
condition to have disobeyed Nature, and the condition of dogs
is such that to obey their natures is to obey us, who have
disobeyed. The implications of this include a vision of the
burden of human responsibility toward at least some animals,
and wishing to be relieved of such a burden is one of the few
reasons I can think of for appealing to a sweepingly mechanistic
view of animal consciousness. Since we are a somewhat lonesome and threatened tribe, I can't think of any other reason we
should desire a philosophy that so emphasizes our solitude.

However that may be, my immediate problems are not
exactly philosophical, although I am driven to philosophy. I
don't, therefore, try to argue against the "wouldn't hurt a
butterfly" bit any more than the "born killer" bit. Winston
Churchill had it that the truth needs a bodyguard of lies, and
I suppose that the more lies, the better in such a situation.

But we are still not out of the woods. Word gets to me of
a plan to poison Belle as a community service. I am not greatly
worried for Belle's sake, since she has already demonstrated her
attitude toward food bribes, and I have systematically encouraged this, but it doesn't help my peace of mind, and I want
more than ever to know something more about what the stories
are about, what we think we can poison off, or maybe sanitize.

We have established that the stories aren't about Pit Bulls, since in order to talk about something you have to know something about it, and even if the logic of horror didn't preclude its being about its putative subject, there would be something almost dishonorable for the active humaniacs about actually checking the situation out. (They would have more trouble getting funding, for one thing.) One way, though, to understand more about the stories is to remember that Bull Terriers were loved and drawn by James Thurber, an ineradicably American artist.

And to remember that they were for a long time America's dog, the most decorated dog of World War I, "The Soldier's Friend," and so on. In 1914 there was a poster by Wallace Robinson that was plainly part of a story America was telling itself about the war situation in Europe. The poster shows, in appropriate national dress, a Russian Wolfhound, a (German) Dachshund, a Pit Bull, an English Bulldog and a French dog. The Pit Bull is in the center, larger than the others, with a gay twinkle in the eye and an American flag around his neck. He is saying, "I'm Neutral but not afraid of any of 'em!" In a tight spot, that's not such a bad story to be telling. The Bull Terrier here, as in many places—Thurber's tales and drawings, or Pete the Pup in the old "Our Gang" series—is an emblem of what it used to be possible to think of as American virtues: independence, ingenuity, cooperation, a certain raffish humor, sticking with your pals when the going gets rough and refusing the aristocratic pseudo-virtues of Europe. "Neutral but not afraid of any of 'em" is central to all of this and is part of the literary tradition that makes Thurber's Rex, who never started fights and never ran away from one, such a meaningful figure.

However you read or value such visions, they have failed, and the stories about Pit Bulls, which got going strong at just the point when it began to look as though the efforts of the as it were freedom fighters of the sixties were going to be bootless, are stories about ourselves, about an America that has gone out of its mind and become in its own visions an unconscious parody of such Bull Terrier-like heroes as Huckleberry Finn. I don't mean that America, or Americans, are writing parodies, but that parody as an art form is well-nigh impossible in a culture that is, or has gone beyond, its own parody.

The stories tell me a lot about how skittish, and dangerously so, America has become, and it is not only the text of the Pit Bull in which this can be read. I am addicted to dog stories of all sorts—the most awful, sentimental child's tale will do. These stories have changed as radically as the stories about Pit Bulls. The older ones were not, mostly, written with Thurber's canny intelligence and humor, but in them there were children, and a dog, and the children learned from the dog's courage, loyalty or wit how to clarify their own stances in the world. In the new sort of story, the initial situation is the same; the dog is for the child the only point of emotional clarity in a shifting world. But the moral of the story is different now. The child may be the son of a violent, alcoholic father and a schizophrenic mother and may fantasize a lot about the glorious adventures of himself and his dog. Halfway through the book, the dog is poisoned, and the boy ends up in Juvenile Hall, where he learns to stop fantasizing and sees that mother and father and teacher were right all along after all.

I sometimes think that the most patriotic thing a movie company could do right now would be to produce a film with a glorious and noble Pit Bull as its hero.

But that isn't happening. Just as "born killer" is symmetrical with, if not identical to, "wouldn't hurt a butterfly," so is the story the White House tells about what a dandy, sweet thing America is symmetrical with the Soviet story about vicious capitalists, and it has the same consequences. Bloodshed and fantods and general misery all around. Horror stories are told to relieve the teller of the burden of judgment—at least, that is certainly a reason I have told various horror stories.

A philosopher recently insisted to me that Wittgenstein was a traitor to philosophy. I had heard that before, but this time I decided to pursue it, to see how the story would go on. It wasn't, he explained, decent or safe to call a question senseless when maybe the question just needed a little refinement. I said that Wittgenstein had provided that refinement. But that wouldn't do; it was wrong of him to have said "senseless" and he couldn't be trusted, that was all.

My response wasn't right, though. Wittgenstein was, at that point as elsewhere, biting. Was he fear-biting or were those precisely administered bites in order? I happen to think the latter, that he loved philosophy as few people can love anything, as a Bull Terrier can love her work, and me, strongly enough to bite.

For example, one day it turned out that three days had gone by and I hadn't worked Belle on retrieving. I was lazing about, reading in bed, on the left side of the bed. Belle brought me her dumbbell and stared at me loudly. (Bull Terriers can stare loudly without making a sound.) I said, "Oh, not now, Belle. In a few minutes." She impatiently dumped the dumbbell on top of the book I was trying to read, put her paws up on the edge of the bed and bit my hand, very precisely. She took the trouble to bite my *right* hand, even though my left hung within easy reach. She bit, that is, the hand with which I throw the dumbbell when we are working. A very gentle bite, I should say, but also just—precise and justified. An inherently excellent moment of exactitude, of the sort Wittgenstein provided out of the harsh depths of love.

All breeds, or almost all, have inherent excellencies. Bull Terriers are creatures that give you the opportunity to know, really to know, should you want so terrible a knowledge, whether or not your relationships and your artistries, your grammars, are coherent, whether what you have is a free-floating and truncated bit of the debris of Romanticism or a discipline that can renew the resources of thought. And, as I have said, unless you are interested in that, get a dog from some other breed, some less deeply domesticated breed.

This is all well and good—but what about dog fights? Oughtn't they to be outlawed and the Pit men jailed? Aside from the fact that it is probably impossible to outlaw dog fights successfully, this question can't be answered without yet more speculation on the nature of things. Also, the question about the cruelty of dog fights is premature in a culture in which debates about dog fighting take place over the lusty pastime of consuming the flesh of animals who have suffered a great deal more than any fighting dog ever does. All I have, anyway, are some guesses about what needs to be brought into the discussion.

Richard Stratton, in various places, argues that dog fighting is not cruel when it is a matter of matching equal dogs against each other in properly regulated matches. He insists (and it is true) that Pit Bulls love fighting contact, and he goes on from there to claim that allowing them to fight is about as cruel as allowing a bird to fly. And even if you aren't swayed by his arguments, when you read his tales of the great pit champions, you do begin to feel that, *with certain dogs*, not only is it not cruel to "roll" them (give them real fighting, as opposed to mere scrapping, experience), it is cruel to prevent them from fighting, in the way it is cruel to put birds in cages, or at least in cages that are too small for them.

But do I know anything about this beyond what I've read and heard? I know it is true that there are dogs who love fighting contact, and that all animals work perfunctorily until they understand the significance and difficulty of what they are doing within the context of whatever love or loves structure the cosmos for them, and that perfunctory work is the death of the soul. There are talents that it is death to hide. Belle retrieves rather sullenly until I start placing her dumbbell in a tangle of vines or under a pile of rocks, making it hard for her, at which point she becomes alive and about twice as big as a constellation. So it is possible for me to contemplate the possibility that allowing the right Pit Bulls, in the hands of the right people, to fight can be called kind because it answers to some energy essential to the creature, and I think of energy, when I think of certain horses, as the need for heroism.

The fights are, unless one dog quits, fights to the death, and I am not ready to put my Belle in such a situation. Besides, if I were to roll Belle and then were to breed her to another Bulldog who had also been rolled, I would have to muzzle both dogs. Otherwise, they wouldn't mate, they would fight; even in full heat a truly game bitch would rather fight than mate.

What sort of perversion of nature is this? someone may want to know. Doesn't this make it obvious that dog fighting is cruel and that it would be a kindness to euthanize all Pit Bulls, including Belle, no matter how good a babysitter she is?

There are a number of large questions that hinge on this, including a complex question about whether the workings of evolution are sufficient bases for moral visions, but the word "obvious" is what has my attention here. A philosopher said to me once in annoyance at yet another tangled, pseudo-Kantian battle about what cruelty is, "Look! Cruelty is not a problem. It is as plain as a pikestaff; cruelty is the infliction of pain unnecessarily, and that is all!"

I wish that was right, that there was no problem, that cruelty was as plain as a pikestaff. If it were, there wouldn't be so much of it around, for I take us to be creatures that quite naturally avoid cruelty—being cruel, I mean—*when we can recognize it.* But we need something else, perhaps a discipline of recognition, for I also take it to be the case that the cosmos is in some disorder, and that this disorder includes our moral perceptions. If the state of things were morally, which is to say philosophically, sound, then cruelty would be obvious. But things-as-they-are aren't philosophically sound. The emblem I offered above, of two dogs who would rather fight than mate, is thus not so clearly an emblem of something *exceptionally* out of order. We do live east of Eden where such an emblem may even be a noble one of caring for what is sacred, what matters, in the way the emblem of two ballet dancers enduring deprivation and pain in order to dance—even in order to dance rather than to mate—is an emblem of something that keeps mattering to human beings.

For fighting-dog people, at least some of them, especially the old-timers, the combination of traits called deep gameness,* which leads to the possibility of dogs who would choose fighting over mating, is in fact emblematic of glory, nobility, dis-

---

*There is, incidentally, evidence that deep gameness has to do with capillaries—it is physiological. That is, a dead-game dog is one that doesn't go into shock. I am indebted to Professor I. Lehr Brisbin for alerting me to this.

cipline in the old sense. In their vision, gameness includes the capacity to choose, and to choose knowingly, nobility and triumph over mere survival—death before dishonor. For me the question, then, is complicated by the fact that there is reason to honor this—it is philosophically much more accurate than the notion of these dogs as mere fighting machines. I have worked and owned dogs who were dedicated fighters, and mortal enemies to boot, who would work together cooperatively on, say, the same relay team with no leashes or any other restraints than their own pride in their work. And I know of English Bull Terriers with significant pit experience ambling and playing together while an evening's fight is being prepared for—and those dogs know what fight preparations are, just as your dog knows what going-for-a-walk preparations are. Not all fighting dogs demonstrate such capacities, which you may want to call inhibitory powers. However, not all baseball players from opposing teams have the capacity to have peaceful meals together during the World Series, either.

Are the Pit men on to something, or is it still obvious that dog fighting is cruel? The word "obvious" has an etymology that gives it the meaning, roughly, of "in the way of the way." This gives me license to suspect that all philosophical truths are obvious in the sense that we can always find a way around them, and that the apparent obviousness of the cruelty of dog fighting is an instance of our having found our way around important aspects of both kindness and cruelty, some of which I indicated in the previous chapter.

Here I should say that, in my parlance, the phrase "fighting-dog people" includes huge numbers like myself who never match their dogs or go to fights and don't want to, but nonetheless value deeply the gameness that gets translated into reliability and hard-going elsewhere, such as still-trailing, scout work or just plain greatness around the house, like Belle's dedicated education of my nephew.

The question is: Where does this leave me with Belle? She shows evidence of gameness, but the only way to be sure of this would be to roll her, and I'm not going to do that, even though I think the argument that rolling fighting dogs is no crueler than taking Greyhounds out for long runs, while it isn't fully sound, also isn't idiotic.

I don't (can't bring myself to) advocate dog fighting. Nor can I bring myself to oppose it. I don't even know what a question about it would look like.

All I can do is brood about the intelligence and moral soundness of the fighting breeds and of the people who love them. That is, all I can do is to construct—with my clues about certain breeders, trainers and owners and their dogs, Belle and her relatives and the astounding greatness of the artist James Thurber—a loose collection of observations, some of which barely add up to so much as a proper *ad hominem*, or *ad canem*, argument. But the light of particular human beings and particular dogs is the only light I know of by means of which to find the beginning of a workable moral philosophy.

10

What It Is about Cats

There used to be, and probably still is, activity in the area called Comparative Psychology that consists of various attempts to work out ways of studying and quantifying memory and intelligence across different species. There was sometimes a certain amount of difficulty in coming up with experimental designs that gave clear results. In one case that I remember something of, various animals were shown the location of hidden food and then brought back minutes, hours or days later and watched to see how well they did in finding the food again. Human beings did moderately well in some of these studies, dogs respectably, but it was the digger wasp that outperformed us all. The way I remember the conversations I used to hear about this, it was less obvious to the researchers than it ought to have been that the digger wasp had shown us that what we call "intelligence" might be a complicated and even chimerical phenomenon. Beyond that it seemed to me as a tracking–dog trainer that not nearly enough had been done to rule out the effects on the tests of the animals' superior abilities, especially scent powers.

I cheered for the digger wasp, because the results in question did at least cause some pause in the machinery of behaviorist speculations. But the animal that defeated such speculations absolutely was the cat. I used to hear older experimenters advising younger ones about working with cats. It seems that under certain circumstances, if you give a cat or cats a problem to solve or a task to perform in order to find food, they work it out pretty quickly, and the graph of their comparative intelligence shows a sharply rising line. But, as I heard, "the trouble is that as soon as they figure out that the researcher or technician *wants* them to push the lever, they stop doing it; some of them will *starve* to death rather than do it." (This violently anti-behaviorist theory never, so far as I know, saw print.)

That result fascinated me—I would have dropped everything in order to find out what the cats were trying to do or say to the researchers. After all, when human beings behave that way, we come up with a pretty fancy catalogue of virtues in order to account for it. But, of course, I was stupidly supposing that the point of these efforts was to understand animals, and it wasn't at all. The point was simply to Do Science, or so I began to suspect when I heard one venerable professor tell a young researcher, "Don't use cats, they'll screw up your data."

What is it about cats? Among gentler and more tentative philosophers than the investigators I describe, cats are considered unobtrusively ubiquitous, and the philosophers are by and large grateful for this. At least, I hear the sound of gratitude in Montaigne when he says to himself that while our way of talking is to say that one plays with one's cat, there is no reason we shouldn't suppose that it is the other way about, that one's cat is playing with one. Montaigne's delicate alertness to such possibilities of grammatical reversal is sadly missing from most modern speculations about language and consciousness, but our cats are still here, which means that the most agreeable of philosophical expressions, the grateful one, is still possible.

The cats who starved to death in the laboratories were, no doubt about it, frustrated animals. The refusal of food is a signal made to the cosmos itself when one despairs of signaling one's chums that something deep in nature is being denied. Infants deprived of touch move in such ways through rage to despair, starvation and death. A mare on the point of foaling will not eat or drink if there is insufficient congruence between her sense of the event she anticipates and the attitudes of the creatures and landscape around her. Children refuse food when they are overloaded with various phoninesses disguised as love, even when they don't go so far as to die. And if you take a house cat and put it in a situation in which there is only one choice, that of responding in a linear way to human expectations, the cat won't eat if eating entails the performance of a kind of "pleasing" that is a violation of the cat's nature, a distortion of the cat's duties on the planet.

This does not mean that cats are perverse, but rather that the pleasures and expectations of human beings are profoundly important to cats. In fact, it suggests that, contrary to popular wisdom, getting it right, accurate, just, about pleasing us is in some ways far more to the point of cat nature than it is to the point of dog nature. Dogs are by and large more like humans in being merely amused and relieved when their imitations and approximations of obedience are accepted by us, and their resemblance to us in that way may be one of the reasons it is easier to achieve general agreement on the interpretation of a given doggy action. But cats take the task of pleasing us far more seriously. Science has shown us this.

Of course, science has also shown us that merely having some lunkheaded expectation and presenting it to the cat doesn't satisfy the cat. The cat's job includes making us aware of the invented nature of our expectations, and cats can't do this when the bulldozer effect takes over our expectations, as it can do in science and in our erotic relationships.

I should interrupt myself and say what I mean by my simpleminded assault on science in general and behaviorism in particular. I don't mean that there is much point in simply discarding, for now at least, such notions as Conditioned Response or Operant Behavior. They are far too useful, philosophically and morally. For one thing, thinking about interactions between stimuli and behaviors without reference to internal events can make it turn out that most things are not our fault, thus relieving us of the "bad conscience" Nietzsche so despised. But there are certain confusions that get into the discussions in practice, usually in the guise of genuine difficulties. The result tends to be that the behaviorist overtly denies the interpretive significance of internal events while covertly making appeal to them when the going gets philosophically rough. The opposite happens too, of course. Some animal trainers declare themselves the enemies of academic psychology without acknowledging the extent to which such things as the Stimulus-Response model has clarified their thinking and practice. All of this is well and good, but it still doesn't turn out that behaviorism in its pure form has come up with a better response to cats' refusals than "Don't use cats, they'll screw up your data."

I am not an especially good observer of cats, so it was a cat who comes when he is called and who performs his interests straightforwardly who first caught my attention. At least, Koshka comes when *I* call him, and he is also tolerably responsive when my queen, Cynthia, hollers at him. He is also somewhat clumsy, which is why it is possible for me to work out fairly easily what he is up to.

Clumsy or not, he is like all cats in his relationship to straight lines. If he is on the windowsill in the living room and I put down a bowl of food on the floor in the kitchen, he selects a route to the bowl that takes him over the sofa and the bookcase and makes it look like a natural route, somewhat in the way a field-trial dog will make his leaps over yawning gullies look natural; it is profoundly important to him that he avoid the stupidities of straight lines. It is because he is clumsy that I was able to see this—the genius of cats is in the way we don't, by and large, think about such things, because they play so sweetly with our expectations, all the while charming us out of false skepticisms.* And they are, as I have said, very serious about this. When they fail at charming us, they move so swiftly

---

*I am aware that a standard explanation of a great deal of cat behavior has reference to mechanisms that are a function of the kind of predator a cat is. This seems to me to be in all sorts of ways a queer sort of "explaining," in part because the model of explanation implicit in such talk demands that we "explain" such phenomena as philosophy with reference to predation (and, of course, some people have done this) because of the analytical focus most higher predators require. In the case of any thoughtful species, it is odd to appeal to events that can be located more or less historically in an evolutionary picture as determining the nature of the present. As if a child's first experiments with finger paints gave us an emblem that "explained" Michelangelo.

to the next meditation that we are hardly aware that there has
been an attempt, much less a failure.

The philosophical condition that makes the cat's indirections
meaningful is one in which we understand that something needs
to be restored, that straight lines, the lines of speech and intention, are already lost to us, which means that our first impulse
toward directness will be irrevocably contaminated. Dogs manifest their sensitivity to that contamination in various ways,
most plainly in their refusals to perform complete retrieves
without the restorations and consolations of formal training,
and cats have their own evasions of post-lapsarian invocations.
One traditional way of understanding Eden has been to say that
it was pre-linguistic, and there is something right about that in
a world in which "linguistic" means "after Babel." But there
is something wrong about it if "pre-linguistic" is understood
to mean "prior to language," for Adam and Eve and God and
all of creation could sing to and call one another. Let us say
that Paradise is not so much prior to language—though it is
certainly prior to our language—as it is prior to epistemology,
prior to doubt about the sources and resources of meaningful
resonances.

In such a case, it is important to understand the circumstances in which cats *will* travel in straight lines and under the
direction of a human. There are people who work cats for
movies. Bill Koehler has had cats whom he could control in
the exacting situations in which the cat's movements must be
coordinated with directors, cameramen, actors and scripts.
These cats are by and large traveling in straight lines in response
to signals (or "discriminatory cues") and for food rewards. Such
cats are spoken of admiringly with such comments as "Open
his cage in the morning and out he comes, jumps on my
shoulder, ready to do a job of work," or "The buzzer sounds
and that cat makes a beeline, right now." That is to say, the
cats are doing in working situations exactly what the researchers
I used to listen to failed to do. The trainers are on to something
that could be expressed by saying that training is partly a
discipline of a kind of negative capability, which they express

in various ways. For example, one day we were watching a woman who was a fine handler work her Basenji on retrieving exercises—and Basenjis are notoriously hard to work with. (I once found myself saying that a masochist is a person who is training his or her second Basenji.) Someone in the group of spectators said, "I like what she does with that dog. Doesn't send out any brain waves." Here "brain waves" was a way of referring to the kind of psychic imperialism I discussed in the chapter on tracking.

By contrast, in the labs where the cats wouldn't eat, I used to see the researcher or the technician or the work-study student walk into the lab, ready to go to work, trying with some degree of sincerity and expertise to be *objective*. This may sound like a corollary of "not sending out brain waves," but in fact it was the first mistake I observed. To be "objective" is to try to approach the condition of being No One in Particular with a View from Nowhere and cats know better than that. They are uneasy around such people because people who don't know better tend to ride roughshod over the cat's own knowledge that a cat is Someone in Particular.

Of course, if the caretaker was an undergraduate, s/he would usually still be moved to talk with the cat, to find the grounds of relationship, but in the laboratory situation the impulse would be truncated, the rhythms of attentiveness and response would be off-beat—and the rhythm and harmony of our attention is everything to a cat. Objectivity depends on models of the world and language that require precisely that flat-footed and contaminated sort of straight line that cats are dedicated to undermining for the sake of clarity and richness of discourse. It has nothing to do with the emptying of self, or really ego, that moves poets to come up with expressions like "negative capability." "Scientific objectivity" is, as most people practice it, precisely what the trainers call "brain waves."

When I was at Gentle Jungle, observing Washoe, there were roughly three categories of people going in and out of the main compound. There was the group that included trainers, handlers and caretakers, there were Hollywood types of one sort and

another and there were academics who were there mostly
because of the presence of the signing chimpanzees. I realized
that I was able, without consciously thinking about it, accurately
and from several hundred yards away to identify which group
anyone who came in belonged to. I wasn't doing this with clues
of clothing either; almost everyone was in the same sort of
jeans, sneakers and T-shirts.

The handlers, I noticed, walked in with a soft, acute, 380°
awareness; they were receptively establishing mute acknowl-
edgments of and relationships with all of the several hundred
pumas, wolves, chimps, spider monkeys and Galápagos tortoises. Their ways of moving *fit* into the spaces shaped by the
animals' awareness.

The Hollywood types moved, of course, with vast indifference to where they were and might as well have been on an
interior set with flats painted with pictures of tortoises or on
the stage of a Las Vegas nightclub. They were psychically
intrusive, and I remembered Dick Koehler saying that you could
count on your thumbs the number of actors, directors and so
on who could actually respond meaningfully to what an animal
was doing.

The academics didn't strut in quite that way, but they were
nonetheless psychically intrusive and failed to radiate the intelligence the handlers did. Their very hip joints articulated the
importance of their theories, they had too many questions, too
many hidden assumptions about their roles as observer. I am
talking about nice, smart people, but good handlers don't
"observe" animals in this way, from within diagrams of the
objective performance, with that stare that makes almost all
animals a bit uneasy, especially cats. ·

Cats do not observe *us* in this way, either—but they do
observe us, almost continuously, as I learned from a poem of
John L'Heureux's, "The Thing about Cats," which closes with
the question:

> A cat is not a conscience; I'm not
> saying that.
> What I'm saying is
> > why are they looking?

It took me some ten years, after being struck by this question, to realize that it was the question I had been looking for, or a real question and a real noticing of the fact that our cats are looking at us. This is evidence of my own participation in the culture's ailurophobia.

I just now looked up from my typewriter at one of my own cats snoozing on top of the stereo. Something—perhaps the longish pause in the sounds of typing—alerted him to the change in my mental posture, and he opened an eye, smoothed a whisker, then leaped down and strolled out of the room with a muffled meow. I felt this to be simultaneously an instance of gracious acknowledgment of the moment of contact with me, together with as gracious a refusal to interrupt me. (I should say that I am quite stern with my cats about their desire to be in between my eyes and whatever piece of paper I am engaged with.)

One could read this small episode in various ways, as mere coincidence or as evidence of my sentimentality about Patrick, but it now occurs to me that the success of language itself may depend a great deal of the time on serendipity, just as it may just turn out that the variations of "meow" that our powers can detect are always, by accident, the right thing to say. Patrick just reentered the room, crossed in front of me with a graceful arch and another unobtrusive comment and settled in a new observation post, in his basket. This felt like the right thing for him to do during another longish pause, during which I muttered aloud, wondering where he had gotten to. It is not in any event a *mistake* on his part, to invoke J. L. Austin's wonderful distinction,* about what remarks and actions of his will fit smoothly into my activities.

But he used to make mistakes. This is not easy to remember, and indeed he so quickly became adept at judging when it was appropriate for him to cuddle, or request a favor, and at what distance from me to be under varying circumstances, that I

---

*In "A Plea for Excuses," in *Animal Thinking* (Cambridge: Harvard University Press, 1984), in which he talks about two instances of shooting a donkey, in one case by accident and in the other case by mistake.

might be forgiven for invoking the notion of an unconsciously "programmed" set of behaviors to account for it. We need a new vocabulary term to notice such errors—a nasty word like "mechanomorphism," for example, or some other way of referring to our thoughtless and superstitious habit of attributing mechanical traits to organisms, as though nature dutifully imitated our inventions. Donald Griffin has pointed out:

> If... an animal thinks about its needs and desires, and about the probable results of alternative actions, fewer and more general instructions are sufficient. Animals with relatively small brains may thus have greater need for simple conscious thinking than those endowed with a kilogram or more of gray matter. Perhaps only we and the whales can afford the luxury of storing detailed behavioral instructions...

But I am in danger here of straying from my investigation. I think that the differences between the case of dogs and the case of cats, and the different superstitious errors we are led into in the different cases, suggest that what we have made mistakes about is the nature of certain virtues, especially the willingness to please. Consider, for example, that there isn't a phenomenon similar enough to ailurophobia for us to have a popular name for it in our relationships with dogs. People just say that they are afraid of dogs, and the fear of dogs is fairly easy to demolish if the right dog and the right handler are about. The fear of dogs usually has a basis that is at least approximately rational, which is one reason why someone who is no longer afraid of Lassie may find their fear reappearing with different dogs or in different circumstances, as in the case of a few friends of mine who are no longer at all nervous about my dogs but are still jumpy when a strange dog goes by on the streets.

Ailurophobia is not like this; it is far more resistant to desensitization techniques (which usually consist of social introductions) and is perhaps more obviously inexplicable. After all, cats are not used much to guard persons or property and are unresponsive to attack training, whereas there are plenty of dogs in the world that are real man-stoppers, as well as quite a few dogs who aren't but who brag that they are when you happen by their yards or their cars. I am, for that matter, sometimes afraid of dogs. That is to say, I respect a dog's assertion of a claim right to property, and in the case of certain dogs I respect their authority when they say, "Do this, not that!"

But the thing about cats is first of all that they are looking at us, and perhaps the thing about ailurophobes is that they don't want to be looked at like that. We are all ailurophobes to the extent that we have bought the culture's "wisdom" about the aloofness and emotional independence of cats, which, as Stanley Cavell has taught me to understand, is logically very like virtually any other expression of skeptical terror about the independent existence of other minds—such as jealousy and the reassurances it demands, or sexism, or racism. So perhaps the aloofness story is one we tell ourselves in order not to know that we are being looked at. But why should we not want to be looked at?

I find that it doesn't help for me to point out that we have various reasons for wanting to hide, if only because that phenomenon has been too often discussed under the heading of pathology, and I am thinking of something that is part of health. And talk in which we say that some people have a fear of intimacy, or that Americans have this fear, or academics have that fear, is similarly unhelpful, as is talk in which we suppose that intimacy as opposed to its false forms is "threatening," or whatever. I don't mean that such ways of talking are wrong, only that what I am interested in is some false ideas we have about the nature of intimacy. The idea, for example, that it consists of reporting on inner states or feelings. This is at best an odd thing for anyone to think in light of the fact that when people are actually spending most or all of their conversational time reporting on their feelings, they are usually boring, and in extreme cases are as likely to end up locked away somewhere as people who seem to lose entirely the capacity to report on their frames of mind when that is appropriate.

There is something that is not *that* that is intimacy. Babies, as Cavell has provocatively reminded us, learn to talk when you talk to them about something—kittens, say, or pumpkins—

and not when they are shut up in boxes, and I want to say somehow that intimacy is thinking. It is thinking about *something*, something other than just the parties engaged in the conversation. If you are my friend, I may from time to time need or want to request your response to some happiness or some grief of mine, and in the logic of any friendship love will entail that we agree to do this for each other. But we won't, as C. S. Lewis observes, want to talk about it once the occasion has passed; it is displeasing to do so unless our interest becomes philosophical. Dwelling on grief and distress or on happiness, or at least dwelling in a housebound way on them, is to dwell in some busy ranch of isolation that is not intimacy and is not thinking. To dwell upon it or in it would be what Lewis, speaking from a precisely British metaphysics of talk, calls an "embarrassment," and what I, speaking from the animal trainer's sense of things, want to call distraction. As when we say, "It's no use trying to talk to her now, she is distracted out of her wits."

I am thinking of the capacity for intimacy as a virtue, the virtue of friendship as what Lewis has called *Philia*, the emblem of which, he says, is two figures holding hands and gazing at some third object. (This is unlike Eros, that love whose emblem is two figures gazing at each other.) Eros may be—in fact had probably better be—figured eventually as the intimacy of friendship. A marriage, for example, may be founded on a rich and continuous conversation about the nature of marriage and love, but it cannot, as the women's magazines keep warning, be founded on continuous *declarations* of love. A friend of mine once said crossly, after a particularly trying evening at another couple's house, that she couldn't imagine herself in a marriage because she didn't feel like spending all of her time praising someone; she preferred thought and conversation. I didn't know enough then to say, "But what that couple was doing tonight— that is not a bad marriage, that isn't marriage."*

---

*Some dogs make continuous declarations of love—or *seem* to—and this can enable some people to survive psychic wildernesses of one sort and another, but it is only training, work, that creates a shared grammar of objects of contem-

Cats do not declare their love much, they enact it by their myriad invocations of our pleasure, and they show their understanding of what they are doing—meaning that they show the structure of their understanding—in part through their willingness to give up the last moment's enactment for this one, as though they knew that love, being what refreshes thought, must itself always be discarding us for our refreshment. You may very well get stuck in yesterday's declarations with your spouse or your child, and they may not know how to prevent this in you or in themselves, but your cat will not permit this. The declarations of five minutes ago, that particular arching of the back, that appealing gesture with the paw, may have been true then but now are not, and the cat never allows them to become the bedraggled hermit in our tropes of gesture "who comes and goes and comes and goes all day."*

One may say with John Hollander that our cats are infinitely interpretable texts, but the "text" is something *between* us and our cats; it is the object the cats make out of our positions relative to each other. We regard it from our viewpoint, they from theirs, which introduces a variation on the theme, as cats seldom want to stand side by side with us holding hands unless they are scared or in certain crises, as our cat Blue was recently, when kitting for the first time. She didn't want my husband to leave the room and even reported to him on her frame of mind and feelings. (But she was asking for this from him, I suspect, because she knows that Robert respects her too much to get stuck in some sticky mode of rescue and can be relied on to go back to the conversation. He doesn't, even while holding hands, send out brain waves. Some people are better at talking with cats than other people are; they have larger capacities for the dreamy yet acute sorts of discourse that most cats seem to favor.)

---

plation outside of the dog and the master, and there where the best conversations start and with them the bonds of that deeper love that consists in thinking.

*From Wallace Stevens' "Notes Toward a Supreme Fiction," in *The Palm at the End of the Mind*, edited by Holly Stevens (New York: Random House, 1967).

Koshka, the cat I spoke of earlier who was clumsy enough
to reveal himself to my blunt perceptions, is somewhat jealous,
or at least he shows his jealousy more obviously than other
cats I have known. He has had to deal, over the years, with a
variety of cats, kittens, dogs, donkeys and other claimants on
the hearth who disturb the progress of his Poem of Koshka.
Nowadays he betrays only the slightest tendency to sulk and
grump, having learned that sulky cats don't please me. But in
his youth this was not so. Once I brought in two fuzzy har-
lequin-marked kittens and sat playing with them on the couch.

Koshka leapt wildly into the middle of this arrangement and
then away, and then back again, screaming hoarsely that it
wasn't *RIGHT*. I batted him in the nose and told him to mind
his manners, but Koshka said shrilly that they weren't minding
their manners, were they? Other, more graceful cats would at
this point have taken to washing their paws perhaps, or have
developed a sudden interest in a squirrel outside of the window
while they worked things out. Koshka retreated to the end of
the couch and looked depressed and forlorn, alternately meow-
ing and purring at me in a loud, unseemly way. Then he
decided to be a good sport and come up and make friends, but
found when he tried to get up that his emotional fit had led
him to getting his claws stuck in the couch (a frequent mishap
for him), and he had to spend a minute or two working them
loose.

Once he got loose, he headed in a straightforward, doglike
manner for me, then seemed to remember himself and went
back to where he had been, lay down again, got up and
zigzagged his way around the room, stopping to sniff some
flowers in a vase, a pile of magazines by the fireplace, until he
finally managed to be hunting an invisible fly that was buzzing
near me and the kittens. Leaping for the fly, he suddenly
"noticed" the kittens and began playing with them, pausing to
rub against me, purring this time in a dignified fashion. This
is not a remarkable cat story, of course.

What had happened so rapidly to transform his clumsily
expressed aggravation into graciousness was, I think, precisely
that typically feline interest in and focus on my pleasure, which

is to say, on my interests—the unstated theme that most of a cat's behavior in relationship to his or her friends is variations on. (This isn't from our point of view an infinitely adjustable pleasing, of course—try keeping cats and parakeets together, for example.) Koshka revealed that theme in his from time to time "doggy" behavior, but most cats don't reveal it so directly, which is why it can seem to us that there is no theme, no focus, to a cat's activities. They stalk the web of our imaginations as carefully as they stalk prey and by and large elude our grosser interpretations with skill and care, not because they wish to remain unknown to us, but because they cannot bear to be *falsely* known, known only by the deceptive glare of a single proposition. Perhaps nervousness about being in such a way falsely known is the healthy source of some of our ill health, our various impulses to hide, to make mysteries of ourselves, as well as the healthy source of ailurophobia. Because cats are more adept than we are at evading monolithic propositions of character, they are also less likely to go insane in the way we do, or dogs and horses do, when "pinned to a proposition."

When our friends get it wrong about us, we tend to go about saying urgently to anyone we can collar that People Don't Understand Us—no one understands scholars, or poets, or animal trainers, or diabetics. And then we go on to try to say what is in fact the case, but one monolithic and totalizing proposition is no better than another, has no more power to penetrate pluralities of perception and misperception.

This, of course, is another error cats avoid. When we aim a misinterpretation at them, they slip sideways so adeptly that it usually seems they just happened accidentally to move at the very moment we took aim and fired, as if it were always by accident rather than by mistake that we miss. (Except, it would appear, when behaviorists get going in laboratories.) Put another way: cats have a much more efficient stroke economy than we do. Here I am using the term "stroke" to mean any stimulus from outside the organism that activates the reticular process (which I heard one psychiatrist call the "starter motor," by way of explaining why strokes are essential to so many organisms).

"Strokes," then, can be any sort of acknowledgment of a creature's existence, and negative ones are effective in at least keeping an organism alive, though unhappy, which is one way of understanding why monkeys will embrace wire mothers and people will stay in relationships that consist largely or wholly of exchanges that leave the participants feeling lousy. Most social animals seem to be capable of becoming addicted to whatever sort of stroke comes handy.

Beyond which, some people seem able to become addicted to "do" strokes rather than "be" strokes, usually in the form of praise for a particular accomplishment rather than for a general way of being. The trouble with "do" strokes is that you can never get enough of them, and their stimulating effect doesn't last very long, hence the dusty trophy cases full of stale strokes that some people clutter up their conversations with.

"Be" strokes, by contrast, can last practically forever and don't require further validation from anyone, including, usually, the creature who gave you the stroke in the first place. So that while I may feel set up for anywhere from a minute to weeks if you tell me that a finished performance is splendid, the thrill will come to an end, whereas if you manage to acknowledge the kind of mind I have accurately, then it is my nature you have acknowledged, something that is by and large immortal so long as I am.

"Be" strokes are the only kind cats are normally interested in, which is why work with them can't go the way work with dogs and horses can. Emotional M&Ms are either ignored or resisted if circumstances make it impossible to perform the preferred feline metaphysics. Hence the grammars of approval and disapproval that so madden humans are refused utterly by cats, who appear to be born with something like an intuitive understanding that approval is almost inevitably the flip side of disapproval, in contrast with some (though not all) dogs, who are like us in that they usually have to spend some time learning the hard way, if they do learn, why it is that bribery and flattery are so dangerous.

Cats' refusal to be approved of or disapproved of may make it appear that, after all, S-R psychology had explanatory force in their case. Especially when people go on to say, as they sometimes do, that in order to get a cat to perform as Bill Koehler does, the "reinforcements" used must be impersonal—the handler's self-esteem must not get into them. Such a way of talking makes tropes of mechanomorphism look philosophically promising. But the advice about the importance of impersonality, like the dog trainer's advice about the impersonality of "Out!" corrections, itself points to fundamental differences between, say, my cat Gumbie and my Jeep Cherokee. My Jeep also "refuses" to run if there is sugar in the gas tank, and so is "finicky." But what the Jeep does that we can call "refusing" is plainly figurative, as is a meter's behavior when we "feed" it. Neither the Jeep nor the meter cares whether or not I care, do not refuse to be "fed" if I make approval noises at them. This sort of difference is so obvious that I am driven to suppose that there must be a very powerful superstition preventing some thinkers from seeing it—thinkers who like to say that a cat cannot be said to be "really" playing with a ball because a cat does not seem to know our grammar of what "playing with" and "ball" are. This sort of more or less positivist position requires a fundamental assumption that "meaning" is a homogeneous, quantifiable thing, and that the universe is dualistic in that there are only two states of meaning in it—significant and insignificant, and further that "significant" means only "significant to me." Such a view demands that we acknowledge that the proposition "Cats are more significant to Vicki than grasshoppers are" is a remark about Vicki, not about cats and grasshoppers in and of themselves, as though Vicki had infinite interpretive powers. Such positivism of meaning looks often enough like an injunction against the pathetic fallacy, but seems to me to be quite the opposite, and also to be, as some writers have claimed it is, a view that does not answer to the theoretical demand for parsimony. If, for example, Gumbie hides when guests she doesn't like come to visit, and stalks about after they leave, suspiciously checking out the evidences of their visit, then my sense of the guests and of Gumbie is revised a bit, especially as Gumbie usually behaves this way when guests attempt uncalled-for familiarity with her, from which it follows that Gumbie is revising the meanings of my world, if I respect her. Of course, I may also say, "Oh, Gumbie, don't be such a snob!" and insist on my earlier, friendlier interpretations of the guests and decide that Gumbie is behaving badly. This will still be a function of Gumbie's interpretive powers, including her power to interpret me, without regard for any theories of Gumbie I may start with. Gumbie may also, while sporting herself in the backyard, so draw my attention to grasshoppers that I become interested in them and maybe take up entomology. If the sentence "Cats are more significant to Vicki than grasshoppers are" is one for which the judgments "true" and "false" are relevant, then it is as much about grasshoppers and cats as it is about Vicki. Compare it with "Xqrwz are more significant to Vicki than bxryqwixxws are." This is not, in the language I speak, anything for which the judgment "true" or "false" is relevant, it is not about anything.

With Gumbie, the only way to manage to believe that any significance she has is the product of my theories about her is to kill her; allow her to live, and she will with every turn, every thoughtful purr and liquidity of comment in her throat, remind me that her relationship to the world is mediated through mine only insofar as that mediation is congruent with the revolving "I Am" that is Gumbie. The objections to my saying this are curiously various. Some philosophers would want, of course, to cry out against my attribution to Gumbie of a concept of self, but others would want to say that the cat's unresponsiveness to emotional bribery is "just" a function of the fact that house cats, like tigers, are loners, not social animals, not dependent on the structure and organization of any sort of group. I don't know where this notion comes from in light of the fact that virtually every popular book on owning cats recommends that you have more than one, so that they will keep each other company when you are not at home.

Cats, unlike horses and dogs, are more likely in domestic situations (hanging around the house) to force the dimmest of us temporarily at least to abandon our epistemological heavy-handedness. When Morris is made out in a TV cat-food commercial to be performing some sort of minuet by means of photographic manipulations, the very ease with which we can so interpret his image is itself a reminder that it is an interpretation built on sand and not a full figuration. We do not forget that "Morris himself" remains outside of our interpretations. Cats are always saying to us in one way and another, "I am the cat who walks by himself, and all places are alike to me," as the cat in Kipling's story does. When a cat looks at us, there is always in the looking the reminder that a cat can look at me or at a king and in both cases equally from the chosen poise of *that* particular angle of grace and speculation.

But, of course, here is the point I am laboring over: They are saying that *to us*. They take infinite trouble that we should continue to be aware of their way of looking. Consider the cat in Kipling. In that story, the first creature to be domesticated was Wild Man, who was "dreadfully wild. He didn't even begin to be tame until he met the Woman and she told him she did not like living in his wild ways." The next animal was of course Wild Dog, who was easily drawn into the amiability of the cave by the woman when she made "the First Singing Magic in the world." The dog came when called and became, by way of a song, First Friend. Wild Horse was cooperative, too, about being charmed and tamed by the Second Singing Magic in the world. And so with Wild Cow. Even the little Bat is a guest rather than an intruder in the cave, and calls the woman "Oh my Hostess and wife of my Host."

But the Cat refused the tale the humans wanted to tell of him and, indeed, insisted on a revision of the woman's story about *herself*, with the result that it was the woman who was charmed, and said, "I knew that I was wise but I did not know that I was beautiful. So I will make a bargain with you. If ever I say one word in your praise, you may come into the Cave." The cat agreed to this and negotiated further for a warm spot by the fire should there be two words in his praise, and the privilege of drinking milk should there be three.

As is usual in such stories, the woman did say three words in his praise, but of course not in the way she meant—the world of such tales is a magically logocentric world in which, as in legal situations, saying "That isn't what I meant!" doesn't get us out of it. The cat, first by tickling and charming Baby, and then by purring and so lulling Baby asleep, finally by catching a mouse, moved the Woman to utter the three words of praise.

Kipling goes on to tell of the return of the man and the dog at the end of the day, and of their threats to throw things and use teeth should the cat fail to continue to be kind to Baby and to catch mice. Kipling falters here, I think, for he has it turn out that the threats are effective against the cat, and I have never seen anyone succeed in making a cat go forward and do something (rather than run away) in response to threats. (In fact, threats aren't really very good motivators for any species that I know of. But that is a somewhat separate issue that has to do with the reasons cruelty doesn't work very well.) What matters here is that, up until the end when Kipling sentimentally allows the dog and the man to succeed with the sort of macho display behavior cats generally despise, he has the important part right, the cat's revisionary impulses.

I don't blame Kipling, of course, for his failure to sustain his cat story properly. It is impossible for a writer to stay ahead of a cat. My cat Blue, for example, is becoming a politician these days and has organized the other cats, who are upset with me because I spend too much time talking on the telephone and making airline reservations instead of paying proper attention to creature comforts. Yesterday a friend of mine called me up because she had suddenly learned more about the nature of the FBI than she had wanted to know, so I started talking to her in an urgent, important-sounding tone of voice. Blue has never particularly liked people whose voices are full of importance, because when people start feeling important, she has to leave off bringing comfort and consolation to my dogs when they are embarrassed that they were goofy enough to bark at the wrong passerby. So, when Blue failed to get my attention away from the telephone, she simply "accidentally" walked on the button that hangs the phone up on her way from the bookcase to her water dish. There was ensuing panic on the part of the party I was talking to, who was quite certain that the CIA or the FBI or someone was trying to prevent us from talking.

By the time I worked out what Blue had been up to, she had given herself a bath, lectured the mice on their behavior and instructed my Pit Bull further on how to keep the male cats in line.

It is pleasing to watch kittens practicing this, stalking a shadow with sideways hops and so on, or playing with parts of their own bodies and those of Mama and their littermates with that odd regard for the intended nature of the tail, paw or ear that makes us tend to say that the kitten chasing his tail doesn't know that it is his tail. What the kitten is born to know is that it is his/her own Tale, the tale of the cat's limitlessly metamorphosing stances toward us and the rest of the world.

I feel again the hot breath of someone wanting to give me a lecture from the opening series in Life Sciences 121a: The Interpretation of Behavior, and tell me that the behaviors I am talking about are explainable as the result of predatory mechanisms in the cat. There is, as usual, an implied "merely" in this, as if in the first place something as difficult and as important as hunting weren't a likely basis for play and tale, and as such also a source of figures of thought in the development of friendships. Believing such a notion consistently would entail denying that any utterance can be a poem, because some or all of its grammar and diction can be shown to have sources in survival modes necessitated by, say, the Pleistocene Drought.

There are differences between the friendship of cats and those of other humans. The cat's insistence on being Herself brings pleasure, whereas such an insistence in human-to-human loves is too often done clumsily and painfully and often results in the static of Quarrel rather than in the Heart of philosophy. But I have been for too long trying to indicate in prose what is more properly celebrated in verse, those turns and graces by means of which not only those who are Beloved Others but philosophy itself consoles us for the very fact of Otherness that drives us to philosophy. Some Eastern thinkers speak of the Gap and then say no more about it. Dogs, people and horses are all likely to try foolishly to close the Gap, to deny our "differences from one another—the one everything the other is not, [to deny] human separation, which can be accepted, and granted, or not. Like the separation from God."* Cats live in a kind of ever-changing song or story in and of the Gap. Here is "Kitty and Bug," by John Hollander:

```
            I       a
          cat     who
          coated in a
          dense shadow
          which I cast
          along myself
           absorb the
           light you
           gaze at me
          with can yet
          look at a king
          and not be seen
           to be seeing any
          more than himself
          a motionless seer
          sovereign of gray
          mirrored invisibly
          in the seeing glass
          of air Whatever I am
          seeing is part of me
          As you see me now my
          vision is wrapped in
          two green hypotheses
          darkness blossoming
           in two unseen eyes
            which pretend to be
             intent on a spot of              bug
            upon
             the
            rug
          Who
          can
          see
            how
              eye
              can
            know
```

*Stanley Cavell, *The Claim of Reason: Wittgenstein, Skepticism, Morality and Tragedy* (New York: Oxford University Press, 1979), p. 496.

11

Rights, Autism and the Rougher Magics

You taught me language; and my profit on't
Is, I know how to curse. The red plague rid you
For learning me your language!

CALIBAN TO MIRANDA in The Tempest

I have been talking up until now about how genuine training disciplines lead to transformations that are psychically miraculous. And the transformation I have been most interested in is the one that takes the trainer and the animal out of the moral life and the comforts of its patent goods into the life of art, a life of uncertain value but characterized by genuine risks and diamond-hard responses and unprecedented responsibilities. That transformation is the one that reveals most fully why any high discipline has from the beginning of Western tradition raised such questions as: "What's it all for?" and "What about Penelope while Ulysses is off being a hero?"

In the lives of young would-be trainers the question comes in the form of: "If you don't start doing your homework and thinking about learning how to get on in the world, we're going to sell that horse!"

The dog and horse stories that help shape and invigorate the trainers' imaginations have also come under fire of this sort.

By "dog and horse stories" I mean a genre that includes *Henry IV, Part I* and *Hard Times*, but for modern readers the most important examples are stories such as *National Velvet* and *Lad of Sunnybank Farm*. Here the confrontations tend to sound like this: It's cruel of you to write that dream-mongering tale that encourages children to yearn for great dogs and grand horses flying down the homestretch instead of teaching them what "reality" is like. In real life such confrontations are by and large pretty stupid and motivated by hidden resentments and jealousies, especially the hidden hope that one's child's life will be a repetition and so an affirmation of one's own, but the questions are nonetheless real. No one knew this better than Shakespeare (who is, incidentally, the poet in whose work is to be found the *locus classicus* of the modern children's horse story) and nowhere is the question about the life of art versus the (mere) moral life more disturbingly raised than in *The Tempest*.

In that play, the figure who represents the artist of great power is Prospero, who has had to flee his dukedom because his devotion to the study of magic has led him to neglect his administrative duties, so he is in exile on an island with his daughter Miranda when the play opens. There are various sprites and other characters on the island who are enchanted by Prospero. One of these is Caliban. He is a violent type, who expresses his rage and resentment of his enchantment by calling it slavery and by continually attempting to rape Miranda and being variously punished for his pains by Prospero.

Miranda, meantime, has patiently taught Caliban to speak—he had no language and virtually no consciousness when Miranda and Prospero first encountered him. She at one point attempts to chastise Caliban by reminding him of the infinite patience she devoted to teaching him language, and he responds not only by saying, "You taught me language; and my profit on't is, I know how to curse," but also by going on with demonstrations that prove that he indeed knows how to curse, as when he says, "The red plague rid you for learning me your language!"

Caliban's learning of language is, for Shakespeare and for Prospero, an emblem of the beginning of art which is also the beginning of sorrow, since through art one learns not only stunning beauties but also the loss of them and therefore grief. Prospero says at one point that grief is "beauty's canker," thus making particular sense of the notion that beauty is the beginning of terror. And one traditional way of reading this play is to claim that it is Shakespeare's farewell to the stage and thus his repudiation of his poetic powers, the play in which the poet's own power is associated with Miranda's ability to teach the savage Caliban to speak. Poetic power is magic power, it has no less than this sort of potential:

> ... I have bedimmed
> The noontide sun, called forth the mutinous winds,
> And 'twixt the green sea and the azured vault
> Set roaring war: to the dread rattling thunder
> Have I given fire, and rifted Jove's stout oak
> With his own bolt; the strong-based promontory
> Have I made shake, and by the spurs plucked up
> The pine and cedar; graves at my command
> Have waked their sleepers, oped, and let 'em forth
> By my so potent art.

For lovers of Shakespeare, it is not extravagant to read this as a claim about the powers of his poetic art, and so when Prospero goes on to say, "But this rough magic / I here abjure... I'll break my staff, / Bury it certain fathoms in the earth, / And deeper than did ever plummet sound / I'll drown my book," it has also seemed reasonable to hear this in Shakespeare's voice. After the end of the play, there is an Epilogue. Some editions of the play say that this is "spoken by Prospero," but others say it is spoken by "The Author"—Shakespeare himself:

> Now my charms are all o'erthrown,
> And what strength I have's mine own,
> Which is most faint: now, 'tis true
> I must be here confined by you,*
> Or sent to Naples. . . .
> . . . Now I want†

*Presumably the audience.

†Meaning: Now I lack.

Spirits to enforce, art to enchant;
And my ending is despair,
Unless I be relieved by prayer,
Which pierces so, that it assaults
Mercy itself, and frees all faults.
As you from crimes would pardoned be,
Let your indulgence set me free.

There is a mood in which it seems reasonable to read the play in this way—the mood of readers or playgoers who have made their own decisions to abjure the rougher magics, that is to say, magics which, while having a will of their own, are solely the responsibility of s/he who releases them and who is thus most vulnerable to them. Rather, the decision is made to live in that realm in which what strength one has is one's own. But there is another mood in which one wants to say that to read these lines as Shakespeare's abjuring of poetry is palpable nonsense. This is the mood of poetry, the mood in which one is when one hears the complex ironies inherent in the sudden appearance at the end of a magnificent poetic performance of what is, by comparison, doggerel. It is, to be sure, very civilized, clever and accomplished doggerel, and in any other context than the one it appears in, surrounded by the soaring poetry of *The Tempest*, it would no doubt be very good poetry indeed. That is part of its painful irony, part of the way the play keeps some of the audience forever hovering on the point of making Prospero's decision without going so far as to do it.

Analogous situations arise in the lives of serious trainers, and there are quite a few people who, learning about the rough magics not only of the high disciplines of training but also of the stories that elaborate the significance of the trainer's discipline, abjure them ahead of time as it were, have sense enough to avoid the tempestuous powers of high art which make castaways of those who encounter them. Some of them become contented pleasure- and trail-horse riders, ignoring the life of the Grand Prix rider, but others try to find an intermediary stance, as serious spectators, neither ignoring nor participating in the exquisite accuracies and dangers. One of the places such people are to be found is in the press room of the National

Horse Show. There are the reporters who follow the great riders and horses all over the world, attending to their progress in minute detail, while themselves remaining safe. They watch Puissance riders and celebrate them, but they watch, as one reporter said, with their hands over their eyes, peeking through their fingers. They can neither abjure nor take on the dangers of Prospero's world.

So I read the closing lines of *The Tempest* not as renouncement but rather as underscoring Shakespeare's terrible awareness, not only that renouncements are possible but that they have their own coherence. I read these lines, then, as someone for whom trail and pleasure riding are not an option. I don't/ can't "go for a ride," I can only work a horse.

The trainer, like the incandescent poet Shakespeare, must at some point confront the ways the horrifying aspects of the Midas story apply to the trainer's condition and worry about whether or not the search for the Golden Age and the Golden Horse doesn't have consequences as terrifying as Midas' golden touch. (The emblem of this question is the horse lamed on a Grand Prix course, or the rider killed on it.) What does the trainer do about the rough magics achieved in actual training, and achieved in part through the rough magics of animal stories?

For a reader or playgoer who is entranced by all of Shakespeare's rough magics and his smooth ones, those lines of the Epilogue scrape across the ears and vision like fingernails on a multitude of blackboards. It is like the discomfort created in the serious trainer watching sloppily run pleasure-horse classes where horses and riders bounce solemnly around the ring while listening to the arguments that go: "But at least these horses, unlike Grand Prix horses and racehorses, are protected from risk, and it gives the horses and riders some healthy outdoor exercise."

So while I think it is in part right to read the closing lines of *The Tempest* as the demonstration of why there might need to be a farewell to art and the reentry of the artist into the moral sphere where sharing different responsibilities is not only accepted but required, there is at the same time the poet's awareness that the lines are a disturbing illustration of what it

would mean to abolish poetry, to bid farewell to the infinite
rigors of the stage, just as a swaybacked, diseased horse was
used by Cervantes to illustrate what can happen when we seek
actually to live a life apotheosizing the divinity of the beauty
of horses but only desultorily, getting it almost right. The case
of Don Quixote, like the case of Midas, also reminds us that
getting it almost right may be the highest (human) possibility,
which is another reason art is so dangerous. To love great
horses or gold is to know how to renounce them, and how to
time it.

Which means for us as trainers that we are allowed to think
of the lines in terms of the trainers' problems about storytelling.
Trainers like the Koehlers have consciously faced this problem
and made their choice. Dick Koehler said to me one day that
while a certain novel that was popular at the time certainly got
closer to the actual way training goes than a story such as *Old
Yeller*, he had decided to cast his fate with *Old Yeller* because
"*Old Yeller* is the truth turned completely upside down, so that
it can be the truth again; that other book, in getting it almost
but not quite right, is evil. That book won't help anyone to
train a dog."

There is an interesting example that shows that Shakespeare
and Prospero and Caliban were literally right about the power
of ordinary language to confront anyone who speaks with the
problem of being human, of having the Midas or Quixotic
impulse. That is the case of autistic children, who either are
born with what appears to be no capacity for language at all
or seem to lose it entirely, in a way dogs never do and of the
success the psychologist Ivar Lovaas has had in teaching such
children to talk.

Caliban is a kind of paradigm case of autism, a condition
that is accompanied by some highly alarming habits, including
"self-stimulation behavior," or "stimming" for short. Stimming
can take a wide variety of forms, anything from a fascination
with running water or colored lights to self-destructive behavior
such as banging one's skull repeatedly against the edges of metal
filing cabinets or chewing one's shoulder to the bone. Autism
is frequently characterized by attacks on other people as well.

This doesn't happen invariably, but it happens often enough so that you can guess the age of an autistic child by the height of the bruises on the parents' bodies.* Ivar Lovaas, the man who has had the greatest success in teaching autistic children to talk and consequently to take care of themselves,† insists that we *owe* it to these children to help them, since they are the ones who pay the price of our evolutionary development—he says, for example, that most of the prized and celebrated behavior that characterizes inmates of universities is a manifestation of autistic stimming, which means that autism is an exaggerated or complete form of those aspects of human intellect and imagination that gives us philosophy, art and Maxwell's equations.

Lovaas' work raises many complicated issues, the most important of which is a question about whether there is any value at all to language and therefore to all of the other things human that depend on language—science, poetry, philosophy, television, love and so on.

This is a question because, in the case of autism, it turns out that while a non-autistic person may suffer and be appalled by the sight of autism, autistic children themselves are apparently quite happy. An autistic infant will lie contentedly for hours in her crib. The mother may suffer from the failure of any relationship to develop, but the baby does not. The capacity for language, for talking, accompanies a capacity to care about whether anyone talks to you; autistic children don't care.

Until they learn to talk. And then they do care. At one seminar, Lovaas showed a film of two brothers, identical twins, both autistic. Lovaas (and his many assistants) use a wholly behaviorist vocabulary and begin teaching language by mechanically shaping it, literally shaping the appropriate physical movements and sounds that go into saying, "Hug me." The

---

*Curiously, the one form of violence that never comes up with autistic children is the one Caliban is most devoted to—rape. Autistic adolescents and adults can live in "homes" with couples who devote themselves to providing these people with something vaguely like "a normal life" and never attack each other sexually.

†Caliban was largely unappreciative of this aspect of his ability to curse.

understanding of what "Hug me" means comes *after* this shaping. In the case of the twin brothers, one of them had learned
the grammar of "Hug me," and the day after he had grasped
it, he went up to his brother and said, "Hug me," holding out
his arms. But the brother was deeply engaged in stimming on
some water running in a sink and didn't respond, and the first
child burst into tears—for the first time in his life.

So the question here posed under the heading "Why learn
language?" is identical to the question "Why be human (what
we mean by human) at all?" In most cases, our humanity is in
place before we can ask the question, because most of us learn
language so quickly and easily that we are already in and of
the problem; autism is not an option. At least, full-time autism
is not. (Lovaas says that the difference between the autistic child
and the professor of psychology is that the professor just doesn't
happen to be autistic all of the time.) We are of necessity driven
to philosophy and perhaps poetry by the paradoxes and muddles
that begin arising the first time saying "Hug me" turns out to
be insufficient to guarantee the response of the other.

But autism is an option for the children Lovaas deals with.
The alternative to the kind of training Lovaas does is a life in
a hospital, continuously drugged and restrained—a life that does
not seem to make autistic people unhappy. They are quite
content. Why interfere with their contentment? Wittgenstein,
who once said, "We like the world because we do," might here
say we do this because we do it. Interfere we must. Why?

Perhaps all of us know what a lie it is that one must suffer
Calvinistically to achieve knowledge of the good and the beautiful, which is to say that what we know is that what we require
for goodness and beauty puts us in great risk of the most
heartrending grief—but always, always, the risk is worth it.
There is nothing behind this knowledge, no reasoning, no
inferring: we are born to it, and our tendency to do something
about autism, like our tendency to imagine and live in the
hardest-edged disciplines, shows how powerful that knowledge
is, which is why states and princedoms dedicate themselves to
educating us out of this knowledge. Boris Pasternak managed

to avoid being so educated, which is why Stalin was afraid of him.

It is in the context of this question that I want to consider again the animal story, the circumstances of its telling and some of the circumstances of its disappearance. I do not want to propose an answer to the specific question about autism because my engagement with that problem is purely as a spectator—I have not earned the right to confront Lovaas and the parents, teachers, trainers and so on who work with him and his methods.

I was standing a few weeks back in the offices of the American Kennel Club. I was there because James Dearinger, for many years the Obedience Chairman and now the Secretary of the AKC, was showing me some of the Club's remarkable collection of paintings of dogs and dog work. At one point he pointed to a portrait of a fawn Boxer and said, "Well, there he is, old Bang'a'way, with that straight front he nearly ruined the Boxers with; all of 'em had that damned straight front for decades, and they're just now starting to recover."

I looked at the noble head of Bang'a'way and said, "Yeah, well, Jim, you know as well as I do that there isn't one of us pure enough of heart to evaluate our own dogs without a real working test."

He said, "That's right, and we are never, any of us, going to be, either, though some of us are better than others."

There is a political, philosophical and literary background to this small conversation, a history of serious thought and living through the question of nobility. Dogs and horses are among the animals that human beings breed not only for their economic value but for sacred reasons. While racing, for example, is largely messed up by human pride and greed, it is nonetheless an expression of our human capacity to value beauty, fleetness and nobility above comfort and convenience. One story about the nature of that valuing is a novel by Arthur Weiss called *O'Kelly's Eclipse.*

The novel is heavily populated by variously noble and igno-ble persons and horses. There are dukes, landed gentry of various kinds; virtually the whole of the English aristocratic system is represented as well as its opponents. Nobility in the sense I have been getting at throughout is exactly the capacity for surrender—to the light, one might say, or to the knowledge I spoke of a few pages back of the *a priori* worthwhileness of genuine risks for the sake of the sovereignty of the good. This is our sacred knowledge but, of course, as Auden says in the remarks I have quoted earlier, that sort of knowledge together with its laws must be embodied in communities and institutions, and there is thus inevitably slippage, and the appearance in the world of travesties of the Good, together with the temptation to suppose that the way to serve the Good is to get rid of the institutions, perhaps to promote anarchy or communism. Sometimes some institutions must be done away with, of course, but there are always new ones in their place, as some people learned the hard way in Soviet Russia by noticing such things as that Marxist prisons were far more horrifying than the Tsarist ones had been. There are a number of mistakes possible here, especially if you have bumped up against false nobility in the form of, say, a snotty English fox-hunting party. One is to confuse power embodied in institutions with the real thing—to think, say, that what goes on behind the windowless walls of the secret societies at Yale is true power and that Yale must be done away with. To think that is to forget that the very corruptions inherent in universities are essential to the structures that house people like Robert Penn Warren, whose nobility is genuine, and give him the time and wherewithal to teach and to write great poetry.

The philosopher Michael Scriven has argued that whatever value system you inherit is likely to be a more flexible and sensitive moral instrument than anything you can replace it with; if you are James Joyce and you inherit Catholicism, you can't simply toss it out. I am saying that we inherit a vision that is embodied in places such as *O'Kelly's Eclipse* and the lives of people who breed and race horses, or run field trials, and that it is as dangerous to throw it out as it was for the Russian revolutionaries simply to throw out Russian aristocratic values.

I must mention one more mistake that can get made, which is to forget that anarchy is probably the most tyrannical institution of all.

Dennis O'Kelly and Eclipse, possibly the greatest racehorse of all time, were historical figures. In the novel, O'Kelly's passion for the "great Harse" is played out during the reign of George III, with Dr. Johnson and Boswell as participants. But what is interesting to me at the moment is not so much what the two of them did or did not do, but their persistence as emblematic figures in the story racing tells about itself, which is why Weiss's novel rather than straight historical chronicle is where I look for an accurate version of the tale.

It is, as almost all stories of great dogs, horses, trainers, owners and breeders are, a story about the mystery of nobility, of finding one's right relationship to it, and especially of what the role of vision and knowledge is. (To anticipate: the racetrack is, like any other institution, a place where the usual course of knowledge is a matter of statistics, of betting the odds. Handicappers, like personnel committees, admissions offices, editors and fellowship committees, are people who are supposed to be good at figuring the odds. Within the world of racing, just as within the world of education, there are people who look at the horses directly instead of at the racing forms, but for most people the racing forms are essential because of the uncertainties of direct evaluation. People who can look at the horses and "pick" them without handicap sheets are the visionaries, the prophetic poets; their powers, like Prospero's, are extraordinary, extra-human, magical, but only from the point of view of people who can't just look at a horse without racing forms and handicap sheets.)

In the story, Dennis O'Kelly is an Irishman and a fugitive from a debased and truncated version of what nobility is. The novel opens with O'Kelly in the midst of this search for the perfect sire and the perfect dam from which to realize his vision of the perfect horse, the horse that will change history, when a hunting party comes his way. He, of course, sides with the rabbit:

Run, bunny, run! I whispered, fading back in the brush.
O these English shamed the harse to use him in pursuit of
blood. Easy it was to hate them whose boot crushed our Irish
necks by the double curse of Cromwell. And even now in
the year of our Cross, a century later, the boot was still on
our throats. This same English boot it was, and fifty blooded
running-harses, that two years ago had sent me in flight for
fear of my life from gasping Ireland.

The low-life and unregarded social position of the visionary,
together with the at times similar position of the animal, is an
archetypal opening for such a story. The dog in "The Bar
Sinister," who brings joy and glory to all concerned, is a stray,
a Bull Terrier that is humbly entered in the great show at the
Garden with "Sire and Dam unknown" written in the appropriate spaces in the entry form. And even in Albert Payson
Terhune's stories about aristocratic Collies, stories that are not
troubled by doubts about the relationship between financial and
social position and moral rectitude, the "lowliness" of the dog
himself—sometimes figured as the dog's literally lower height—
is essential to the working out of true moral and artistic vision,
courage, intelligence and so on.

In such stories, this requirement is not a function of the
automatic corruption inherent in lofty status but rather of defects
in human perceptions of the noble and true. In fact, O'Kelly
himself is of noble blood, and his character as a great horseman
is *inherited*, is a birthright:

> . . . the Irish nation starved except for a fortunate few among
> whom were the Carlow O'Kellys.
> At Tullow, in the County of Carlow not far from the
> Curragh O'Kildare where the running-harses ran, the O'Kelly
> larder was never empty, for my father was breeder of blooded
> harses and farrier to English gentlemen. It was the limestone
> subsoil, he said, which made Ireland the best place in the
> world for the raising of God's chosen animal. It must have
> been so, for he knew all there was to know about these lovely
> creatures of which more than a hundred were in his care.

And when O'Kelly's use of fifty of his father's horses to level
hated English fences leads to his having to flee Ireland, he

refuses to take what seemed to be the sensible alternative of going to America in these terms:

> Run where? Across the seas to Maryland plantations where already American Lads were grumbling as loud as Irishmen? To Georgia to live with convicts?... The West Indies to die of fever, New York of cold? Run? No, not I! Dennis O'Kelly was not born to die before his time, nor thirst while others drank, or stand while others danced. No, nor wear homespun while others wore brocade!...
>   I'll go...To England! Where station can be won....By what I know of harses!

This he does. The novel is an adult version of the classic children's story of the human character who fights his or her way victoriously through a web of the paradoxes of knowledge and acknowledgment that is as splendidly structured as any allegory I know. At the end it is a woman, Kate Warren, who has ridden Eclipse in the heat that leads to the famous cry "Eclipse first, the rest nowhere!"* So part of O'Kelly's task, like the tasks of Launcelot and Percival, has been getting it straight about women.

When the novel is near its close, O'Kelly has not stopped grumbling about the English boot, despite the powerful help he gets from people like Boswell. After Eclipse's astonishing victory in the qualifying heat, the Judge of the Course reverses an earlier decision forbidding O'Kelly to ride, and then, one by one, the other owners withdraw their horses. This was quite a race—most of these were horses whose names have been immortalized even outside of racing by George Stubbs: Gimcrack, Luster and so on—but O'Kelly is as usual too busy raging against the English to think through the significance of this until the very last moment, when the Duke of Grafton, "whose name would go down in history as a Prime Minister of England," withdraws and says:

---

*This phrase isn't vaguely but rather exactly descriptive of the race—Eclipse crosses the finish line before any of the others have passed the distance post.

"Englishmen all! I speak for all of the gentlemen whose horses were to run this race. But there are times to race and times when an Englishman must withdraw."

Although Kate held tight to me, I could still feel the threat of her tears in the tremble of her hand. I whispered fiercely:

"Kate, we'll stand against them. We'll stand against the world together!"

"We have withdrawn," the Duke continued, "for by so doing we signalize our recognition that a new breed of running-harse has been brought to light. And by our withdrawal we bow to Eclipse... and to Mr. O'Kelly."

And here the novel ends, since there is nothing else that can happen—except for Eclipse and O'Kelly to make their victory walk across the finish line.

In this book, the difficulty of being true to one's own vision is a persistent problem. At one point, for example, O'Kelly has become the lover/slave of Lady Leila while he is learning the manners of English society and placing the wagers that will give him the money he needs to buy the dam and sire of Eclipse. O'Kelly stays with her, not for the entry into society she can give him, but because she owns Spileta, the dam he wants.

His vision seems to fail him. Before he took up with Lady Leila, he could go to the races and listen to the horses, who would whisper to him when they were going to run and win. Now they still seem to whisper, but the ones he bets on don't win, over and over again. This he attributes to his having accepted her enslavement of him, to the gilt snuffboxes, the wearing of brocade, and thus to the corruption of his soul—he is a kind of Launcelot whose loss of chastity signals the failure of his quest for the Grail. But nothing of the sort is going on. It is simply that Leila is fixing the races so that he loses his money; she is using ordinary criminality to attempt to ensure his dependence on her, not some mysteriously corrupting erotic power.

In this relationship and in other ways, O'Kelly learns to evaluate English culture correctly. He fights for John Wilkes, is taught to read literature by Boswell. And, like Spenser's knights, he gets it all wrong at least once in every situation that confronts him; his closing mistake, in interpreting the withdrawal of the other horses as a refusal on the part of English aristocracy to meet an Irishman in a fair race, is typical.

It is important to know what kind of hero O'Kelly is. He is, to be sure, a great breeder, and that greatness is connected to his ability to know horses directly, with a knowledge unmediated by statistics and pedigrees. Indeed, when he recognizes in Spileta the dam he has been looking for, he does it in the dark, by feel. This means that his knowledge, his gift, is not the usual gift of responsibility and commitment that merely good breeders require. His "strange brotherhood" with horses is his knowledge, and the commitments having such a knowledge entail are greater than the burdens imposed by the ordinary demands of great breeding and training.

Few breeders or trainers, whether gifted or not, find themselves in the kind of historical circumstances Weiss puts O'Kelly in, in which large significances and energies gather at every turn. For the ordinary audience, the story is an allegory of what correct behavior would be for those who might find themselves in the vicinity of an O'Kelly, so it is a tale for most about learning how to know O'Kelly, rather than how to be him, how to have his knowledge. It is James Boswell who is the emblem of the great gentleman. In the first pages of the novel, he sees and speaks with O'Kelly at Epsom Downs, recognizing instantly true genius in this unknown Irishman. But even Boswell is not the model, for he, by his own account, is also in a special sense beyond the ordinary, as he has a gift to live with and up to, a "genius for knowing genius," as he says. The other horse owners in the famous race, withdrawing their horses, are the models, perhaps, for most readers, of how to find their own nobility in a confused and impertinent world.

The kind of story I am talking about does not generally recommend the knowledge of and commitment to animal nobility as a way of life to anyone who is not already blessed or cursed with a mind and heart given over to animals. In, for example, *Algonquin: The Story of a Great Dog*, there are two brothers, Uncle Ovid and Grandsir, and the narrator is a boy who is their great-nephew and great-grandson. It is Uncle Ovid, the ne'er-do-well brother, who trains and shows Algonquin in field trials, thus establishing the dog's nobility. The pup was originally to have been a gift for the boy, but such a dog demands responses—the assumption of responsibilities—beyond the humanly demandable:

> Mostly what I went back there for was the puppy, to get him and take him home to be my dog, the way a boy wants a dog. But when I saw him and saw the ruins, I knew all at once how ridiculous it was. He had not been saved from the fire to be a playmate for a little boy. Not for that, Indian Maid, with her hair burned off to her skin and the skin black where it was not burning with the oils in it, moving relentlessly through the flames because she could not die until she was done. Not for me.

The fate of the trainer of vision, Uncle Ovid in this story, is at once a higher fate and a lower fate than that of the Judge, his brother. Grandma explains that Ovid had the same abilities his brother had, but he never amounted to anything because he was always "going out to see to something about the pups."

In this book, the way of life Ovid chooses commits the quester to a sustained knowledge of death of a particular sort, a knowledge the Judge has decided against living with, and he admonishes Ovid angrily at one point, saying, "Don't you know how much of your life you put into every dog, and how much you are diminished when each one dies?"

"I don't want to know," Uncle Ovid said. "I'm an old man, and it wouldn't be good for me to know."

I have here only very sketchily indicated the nature of the stories I am talking about, the background to my conversation at the AKC with James Dearinger. What these stories are about is what it means to have something at stake, how to be able to do that, and what, as well, language and the imagination have to do with it. In these stories the logic of nobility is the logic of language. In *O'Kelly's Eclipse*, for example, the coupling of Dr. Johnson's quest for a great dictionary and O'Kelly's for the perfect horse is no accident, nor is it accidental that O'Kelly is the one with the power to revise Dr. Johnson's dictionary, in a conversation about Johnson's famous misdefinition of the word "pastern." And it is not an accident that O'Kelly needs instruction from Boswell in reading Shakespeare as preparation for his task. In McKinlay Kantor's *The Voice of Bugle Anne*, it is talking to the dog, listening for the dog, telling tales about the dog and even doing murder—not *for* the dog herself, but rather in response to the inexorable "logic of the inheritance that put that bugle in the hound bitch's throat"—that reveal the central importance of telling. The cry of the great foxhound defines imaginative possibility, as it also does in Harriet Arnow's stunning novel *Hunter's Horn*, where, as in *Algonquin*, everything hangs on the decision to answer the cry of the great hound. In these books, one can see with terrifying (if one wants to be terrified) clarity what the decision to honor the music of the hounds and the tales and paintings of the great ancestors of Algonquin amounts to, how someone like Ovid seems to be choosing that tradition as well as the greatness of a single dog over everything else that one might call civil or domestic value. These books relentlessly press the question "What's it all for?" in terms of a question about the value of the human imagination, and specifically about the literary imagination, by which I mean not only what gets written down in books and entered into the official canon of literature, but also all of the large and small tales we tell ourselves about who we are and what we will do. When the characters in *Hunter's Horn* confront the obsessed owner of the great foxhound, it is not the hound herself they would have him relinquish, but rather the *story* he tells about the value and greatness of fox hunting. For him, of course, to give up the allegory of fox hunting is equal to giving up the hound.

In the stories, the characters who decide against the heritage of the imagination do so for the same reason that Prospero abjures his magic and drowns his book. These are characters who begin in pursuit of something larger, more terrible and wonderful than ordinary human happiness, just as Lovaas, in teaching autistic children to talk, is moving them into a realm that appears to be beyond their happiness. The autistic child, learning to speak, learns to desire another because that is in the grammar, and learns grief thereby. Prospero enslaves Ariel and Caliban for his magic and neglects the affairs and concerns proper to a duke. Ovid is dependent on his more cautious brother since you don't earn much of a living if you're out in the kennel and field all of the time. In *Hunter's Horn*, a father has—unwittingly but inevitably—sacrificed his daughter's happiness to his love of the quest. O'Kelly stands willing to sacrifice his own and Kate's chances for ordinary human love to his search for the perfect racehorse. These are Mephistophelean stories in which the possibility that what we often call the imagination, and what is at other times called knowledge, is meticulously examined as the possible source of evil in human affairs.

In many of these stories, the central character is one who turns down the quest. Prospero drowns his book and gives us polite verses at the end of *The Tempest*; the boy in *Algonquin*, when he is offered another dog by Grandsir, thinks about it and says that he might like a nice Irish Setter because you're not likely to run into a great dog if you go for Irish Setters. Like Grandsir, he is unwilling to take on "the grief and the rage and the ashes." This is also the point of view of the wife in *Hunter's Horn*, who suffers loneliness and hardship, as do her children, because of her husband's dreams.

Is disappointment the revelation in these books? Did Shakespeare live only to show the failure of poetry to be worthwhile? Must the hero give up his horse if he wants to get the girl and keep her happy? I don't think so; these stories, like *The Tempest*, reveal the nature of what may be equally called a blessing or a curse, as the narrator of Isaac Taintor Foote's splendid tale *The Look of Eagles* at one point describes "a life and a heart given over to horses."

But the writers and publishers of children's books in this country seem to have decided, as did the educational philosophers of the nineteenth century satirized by Dickens in *Hard Times*,* that a life and a heart given over to any realm of the

---

*A novel in which performance dogs and horses, highly trained ones, that is, represent the life of fancy and the only hope for the human characters in the book.

imagination is unequivocally a curse. The children's dog and horse story that encourages such lives and hearts as were encouraged by *National Velvet* is vanishing rapidly, replaced by a kind of story I have written about earlier in this book, a story in which the child's love of animals and imaginative response to them prove to be insufficient and even antisocial, and must be abandoned. I haven't yet worked out just when this new sort of story replaced the kind I am talking about (a kind that is, by the way, not only accidentally Shakespearean in its visions, but specifically in some cases the literary heir to the characters of horses and horsemen in Shakespeare), but I suppose it happened in the sixties, at about the time some of my friends and colleagues started saying lugubriously, "The notion of heroism is bankrupt in our time." Someone may argue that an earlier book like *The Yearling*, in which the deer must be killed at the end in order that the family can survive, is a book in which children are encouraged to forget dream-mongering and grow up, but that book is very different from the newer kind; there, as in *O'Kelly's Eclipse* and all of the books I have been talking about here, the central revelation is of a perfected form of beauty and of the ways learning about beauty, learning to be fully human, entails honoring it while at the same time remembering that grief is "beauty's canker." The newer kind of story doesn't allow the reader to know that there can be such beauty and has forgotten how to show what knowing that "beauty is the beginning of terror" can command of us.

The new stories also fail because the authors and editors seem to have forgotten, or never to have known, what Wittgenstein had in mind when he said, "To imagine a language is to imagine a form of life." It is a characteristic not only of great trainers but of great animal stories that they honor as fully as they can the actual language of training. (Just as Shakespeare did, not only with horsemanship but with everything human that his pen touched.) There are still a few such books, such as *Nop's Trials*, but these are books about adults, and written by and large for adults. The new children's stories tell the young reader that attempting to reclaim the biting Doberman is not only hopeless but likely to give Mother an anxiety attack,

that to want a horse passionately is to betray the branch of the
Jewish family that is still behind the Iron Curtain, that to love
a Staffordshire Terrier is to encourage dog fighting, rape, murder and blackmail. They are written by people who do not
know what they are talking about, whose only motive seems
to be teaching modes of pity and guilt instead of modes of
agape, or modes of love and heroism. *The Yearling* moves one
to grown-up pity and terror and love; the new story moves
one to false pity, guilt and dejection.*

I suspect strongly that these new stories, which appear to
be honoring animals by elaborating on their suffering, are a
response to what Cavell calls our "skeptical terror" in the face
of the Otherness of anyone or anything that is Beloved. I also
believe that to be fully human is to recognize everyone and
everything in the universe as both Other and Beloved, and that
this entails granting that the world is authentic and meaningful
without demanding proof, without insisting on the kind of
doubting I spoke of that goes on in the discussions of the
signing apes. To tell stories about the pitiful plight of animals
is to deny their Otherness, their autonomy, and thus to deprive
us of one of the central benefits, for humans, of the fact that
animals exist. Everything in the universe is, as I have said,
Other, but animals are the only non-human Others who answer
us without our having to travel to India and find the right
guru.

Trees are of a benevolence that gives them standing, and
that benevolence has even inspired philosophers to speculate
about whether or not trees have rights. Nonetheless, trees don't
answer in the ways animals do, and most of animate creation
doesn't answer as loyally and with as much respect for the
details of the human landscape as dogs, cats and horses do. It

---

*The reader may want me to name the corrupted stories I am referring to.
I don't see any point in this, though it might be interesting to a sociologir .. I
have had, more than once, the experience of being approached about doing scripts
for animal movies or writing children's animal stories. And the requirements
are quite specific: Trainers must be shown to be cruel and animals must be shown
to be pitiful, and life must be shown to be largely a matter of diminishment and
dejection. Joy and commitment to anything but guilt are suspect.

is then the sacredness of answering, for a tribe as lonesome and threatened most of the time as ours is, that makes animals matter. And I suspect knowing that speaking and answering are sacred is the foundational justification for Lovaas' work with autistic children—who are Others who *don't* answer without extraordinary efforts on our part.

The fact that animals are so generous in answering us is what makes it not only okay to train them but a human duty, one way we enact our gratitude to the universe that animals exist. It also makes the telling of animal stories that are the genesis of the courage of the trainer part of our duty, our duty to have the imaginative courage to persist in telling our favorite dog and horse stories against the blunt and blundering assaults of academic psychology and philosophy. (Academic psychology's role in this situation was, in the past, played by the Church. In eighteenth-century France the Jesuit philosophers Péres Bougeant and Gresset were excommunicated for writing books, and in Gresset's case a poem about a parrot, in which they made arguments not unlike the ones I have been making in this book. The nature of the offense of anthropomorphism—heresy—has not changed, only the institutions that perform the excommunications.)

So I am ending my book by appealing to the sense I have developed, as a result of reading and thinking like a dog and horse trainer for several decades now, that animals matter to us, and that the way they matter to us is probably all we can know of how and why we matter and of how they matter to one another and to the planet. The animal trainer's version of Genesis will and must continue to be the one I told earlier, the one that ends with a picture of Adam and Eve leaving Eden accompanied by the few species who chose to share their lot, to accept the human fate and all of the uneasiness and dis-eases that implies.

My Pit Bull is lying on the floor beside me, waiting uncertainly for me to leave off writing this book and attend to her. The knowledge in that waiting of hers—a knowledge I can never know completely—is the gift I am speaking of. Knowledge of animals and animal training is not the only example of the largesse of the gods, but it is one of the most important, given, as Xenophon says, "to the heroes, for their greater knowledge and glory." And to wrench out of context some lines from a lovely poem of James Merrill's: "About the ancient bond between their kind and ours, little more to speak of can be done."* This means, of course, that we need heroes, and we need stories about heroes, as part of our way of ensuring that animals are not denied their fundamental right. This is a right that can be and indeed is violated continually, but it is a right that for all of that can't be anything but unalienable. This is the first right, the right from which all others follow, for them and for us, the right to be believed in, a philosophical right to freedom of speech, the right to say things the philosopher has not taught them or us how to say.

*The poem is "In Monument Valley," and it appears in *Braving the Elements* (New York: Atheneum, 1972) as well as in the more recent *From the First Nine* (New York: Atheneum, 1983). The poem is about driving through Monument Valley in a Hertz car and coming on a starving horse, a mare, and being thereby reminded of an earlier ride on a "buoyant sorrel mare." The speaker offers the mare his apple core, but she refuses it, "lets it roll in the sand." The driver rolls up his car window and observes, "About the ancient bond between her kind and ours, little more to speak of can be done."

Afterword

This is the wrong book in which to discuss this in detail, but a step toward granting the kind of rights I speak of here would be to pass legislation, a Companion Dog law, that would grant to dogs whose owners have put in the time to get real off-lead control the same privileges now granted to the dogs who work with the blind. The discussion of what this would mean in detail is technical in a way this book is not, but I'd be glad to work with any clubs or legal groups on the drafting and instituting of such a law.

Index

A NOTE ABOUT THE AUTHOR

Vicki Hearne (1946-2001) was born in Austin, Texas, graduated from the University of California, Riverside, and became a professional dog and horse trainer in the Riverside area. She trained dogs for obedience classes, the hunt field and scenting work, and trained owners to train their dogs. She also obedience trained several wolves and one goat. She trained horses for the show ring, Grand Prix exhibition and the racetrack. In 1976, she won a Stegner Fellowship to Stanford University and did graduate work there. From 1980 to 1984, she was a lecturer in creative writing at Riverside. In 1984, she went to Yale. She published two volumes of poetry, *Nervous Horses* and *In the Absence of Horses*, and many magazine and newspaper articles.